T0214507

Lecture Notes in Bioinformatics 11371

Subseries of Lecture Notes in Computer Science

More information about this series at http://www.springer.com/series/5381

Sören Auer · Maria-Esther Vidal (Eds.)

Data Integration in the Life Sciences

13th International Conference, DILS 2018
Hannover, Germany, November 20–21, 2018
Proceedings

 Springer

Editors
Sören Auer
TIB and Leibniz University
Hannover, Germany

Maria-Esther Vidal ⓘ
TIB and Leibniz University
Hannover, Germany

ISSN 0302-9743   ISSN 1611-3349   (electronic)
Lecture Notes in Bioinformatics
ISBN 978-3-030-06015-2   ISBN 978-3-030-06016-9   (eBook)
https://doi.org/10.1007/978-3-030-06016-9

Library of Congress Control Number: 2018964607

LNCS Sublibrary: SL8 – Bioinformatics

This Springer imprint is published by the registered company Springer Nature Switzerland AG
The registered company address is: Gewerbestrasse 11, 6330 Cham, Switzerland

Preface

This volume comprises the proceedings of the 13th International Conference on Data Integration in the Life Sciences (DILS 2018), held in Hannover (Germany) during November 20–21, 2018. DILS 2018 was hosted by TIB Leibniz Information Centre for Science and Technology, L3S Research Center at Leibniz University of Hannover, the Information Centre for Life Sciences (ZB MED) and the Hannover Medical School (MHH).

The articles included in this volume went through a peer-review process where each submission was reviewed by at least three reviewers and one senior program chair. The submissions were evaluated in terms of relevance, novelty, significance, soundness, and quality of the presentation. Three types of submissions were received: (1) full papers describing solid and complete research contributions; (2) short papers presenting results of on-going research work; and (3) poster and demonstration papers. We accepted five full papers; eight short papers; four poster papers; and four demo papers. Our sincere thanks go to the Program Committee members and external reviewers for their valuable input, and for accepting our invitation to contribute to the review process.

The DILS 2018 submissions cover a wide variety of topics related to data management in the life sciences. The articles tackle open problems and technical solutions for data integration, query processing, and analytics on big life science data coming from diverse data sources, e.g., genomic data collections, biomedical literature, or clinical records, and in the challenges of transforming big data into actionable insights.

We composed an exciting program that included four research sessions: (1) Big Biomedical Data Integration and Management; (2) Data Exploration in the Life Sciences; (3) Biomedical Data Analytics; and (4) Big Biomedical Applications. Additionally, the program included two invited talks; the first invited talk was on "Matching Biomedical Ontologies for Semantic Data Integration" by Dr. Catia Pesquita, and the second invited talk was on "The de.NBI network–A Bioinformatics Infrastructure in Germany for Handling Big Data in Life Sciences" by Prof. Alfred Pühler. Posters and demos were presented in a plenary session where the authors and attendees had the opportunity to interact in an informal environment.

November 2018

Sören Auer
Maria-Esther Vidal

Organization

General Chair

Sören Auer · TIB Leibniz Information Centre for Science and Technology, and L3S Research Center at Leibniz University of Hannover, Germany

Program Chair

Maria-Esther Vidal · TIB Leibniz Information Centre for Science and Technology, and L3S Research Center at Leibniz University of Hannover, Germany

Organizing Co-chairs

Thomas Illig · Hannover Medical School (MHH), Germany
Wolfgang Nejdl · L3S Research Center at Leibniz University of Hannover, Germany
Dietrich Nelle · ZB-MED Information Centre for Life Sciences, Germany

Local Organizers

Alexandra Garatzogianni · TIB Leibniz Information Centre for Science and Technology, and L3S Research Center at the Leibniz University of Hannover, Germany
Katrin Hanebutt · TIB Leibniz Information Centre for Science and Technology, Germany

Program Committee

Maribel Acosta · AIFB, Karlsruhe Institute of Technology, Germany
José Luis Ambite · University of Southern California, USA
Naveen Ashish · InferLink Corporation, USA
Diego Collarana · University of Bonn, Germany
Benjamin Lang · Centre for Genomic Regulation (CRG), Spain
Marcos Da Silveira · Luxembourg Institute of Science and Technology, Luxembourg
Michel Dumontier · Maastricht University, The Netherlands
Kemele Endris · L3S Research Centre at the Leibniz University of Hannover, Germany
Juliane Fluck · Fraunhofer SCAI, Germany

Konrad Förstner	University of Würzburg, Germany
Mikhail Galkin	University of Bonn, Germany
Martin Gaedke	Chemnitz University of Technology, Germany
Matthias Gietzelt	Hannover Medical School (MHH), Germany
Irlan Grangel	University of Bonn, Germany
Anika Groß	University of Leipzig, Germany
Udo Hahn	Jena University, Germany
Robert Hoehndorf	King Abdullah University of Science and Technology, Saudi Arabia
Patrick Lambrix	Linköping University, Sweden
Pierre Larmande	Institute of Research for Development, France
Ulf Leser	Humboldt-Universität zu Berlin, Germany
Angeli Möller	Bayer Business Services, Germany
Bernd Müller	ZB MED Leibniz Information Centre for Life Sciences, Germany
Isaiah Mulang	University of Bonn, Germany
Guillermo Palma	L3S Research Centre at the Leibniz University of Hannover, Germany
Catia Pesquita	University of Lisbon, Portugal
Cédric Pruski	Luxembourg Institute of Science and Technology, Luxembourg
Erhard Rahm	University of Leipzig, Germany
Alejandro Rodriguez	Universidad Politécnica de Madrid, Spain
Kuldeep Singh	Fraunhofer IAIS, Germany
Dietrich Rebholz-Schuhmann	Information Centre for Life Sciences (ZB MED), Germany
Andreas Thor	University of Applied Sciences for Telecommunications Leipzig, Germany
Johanna Völker	Bayer Business Services, Germany
Amrapali Zaveri	Maastricht University, The Netherlands

Sponsoring Institutions

TIB Leibniz Information Centre for Science and Technology, Germany

Contents

Big Biomedical Data Integration and Management

Do Scaling Algorithms Preserve Word2Vec Semantics? A Case Study for Medical Entities

Janus Wawrzinek$^{(\boxtimes)}$ ⓘ, José María González Pinto ⓘ,
Philipp Markiewka ⓘ, and Wolf-Tilo Balke ⓘ

IFIS TU-Braunschweig, Mühlenpfordstrasse 23, 38106 Brunswick, Germany
{wawrzinek,pinto,balke}@ifis.cs.tu-bs.de,
p.markiewka@tu-braunschweig.de

Abstract. The exponential increase of scientific publications in the bio-medical field challenges access to scientific information, which primarily is encoded by *semantic relationships* between medical *entities,* such as active ingredients, diseases, or genes. Neural language models, such as Word2Vec, offer new ways of automatically learning semantically meaningful entity relationships even from large text corpora. They offer high scalability and deliver better accuracy than comparable approaches. Still, first the models have to be tuned by testing different training parameters. Arguably, the most critical parameter is the *number of training dimensions* for the neural network training and testing individually different numbers of dimensions is time-consuming. It usually takes hours or even days per training iteration on large corpora. In this paper we show a more efficient way to determine the optimal number of dimensions concerning quality measures such as precision/recall. We show that the quality of results gained using simpler and easier to compute scaling approaches like MDS or PCA correlates strongly with the expected quality when using the same number of Word2Vec training dimensions. This has even more impact if after initial Word2Vec training only a limited number of entities and their respective relations are of interest.

Keywords: Information extraction · Neural language models
Scaling approaches

1 Introduction

The current exponential growth of scientific publications in the medical field requires innovative methods to structure the information space for important medical entities, such as active ingredients, diseases, genes, and their relationships to each other. For instance, a term-based search for a common disease such as diabetes in the medical digital library PubMed leads to a search result of over 600,000 publications. Here the automated extraction of high-quality relationships between entities contained in medical literature would provide a useful tool to facilitate an exploration of large datasets. Moreover, such an extraction could serve as a basis for numerous innovative medical applications such as Drug Repurposing [2, 6], the discovery of drug-drug interactions [3], the creation of biomedical databases [4], and many more. Previous work has

© Springer Nature Switzerland AG 2019
S. Auer and M.-E. Vidal (Eds.): DILS 2018, LNBI 11371, pp. 3–16, 2019.
https://doi.org/10.1007/978-3-030-06016-9_1

recognized this and proposed several methods to calculate similarities between entities to infer their relationships. These include popular approaches such as the computation of chemical (sub-)structure similarity based on bit-fingerprints [7] or methods relying on entity networks [5]. Recent approaches even try to calculate similarity based on word contexts using distributional semantic models (DSMs) [1, 3, 8, 9]: here, a similar word context points to an implicitly expressed relationship. This property is often transferred to entities: two entities in similar linguistic contexts point to an intrinsic relationship between these entities and possibly also to similar entity properties. According to Baroni et al. [10], DSMs can generally be divided into count-models and predictive models. For count-models, first word-context matrices are generated from a text corpus, followed by matrix optimization steps such as re-weighting and dimensional scaling [10]. In contrast, predictive models (also known as embedding models or neural language models) try to predict an expected context based on numerous training examples. Studies show that state-of-the-art predictive models, such as Word2Vec, outperform count models in performance and scalability, in particular in semantics and analogy tasks [10, 11].

Recently researchers [26–28] have tried to uncover the theoretical principles of Word2Vec to reveal what is behind the embedding vectors' semantics. In particular, the work of [28] has demonstrated that a reformulation of the objective of the skip-gram negative sampling implementation (SGNS) of Word2Vec leads to a mathematical demonstration that SGNS is, in fact, an explicit matrix factorization, where the matrix to be factorized is the co-occurrence matrix. However, little is known about the effect of scaling algorithms on Word2Vec: do we lose its appealing semantics, or do we filter out noise [17]? Among the popular scaling algorithms that exist, which one can preserve the original semantics better? Does it make a difference which scaling algorithm is chosen? Answering these questions can help researchers to find the optimal number of dimensions of semantic spaces efficiently. In fact, the usually accepted '200–400 dimensions' chosen when training Word2Vec (see e.g., [11, 12]) has yet to spark a more in-depth investigation.

In this paper, we pragmatically investigate these questions to provide first insights into the fundamental issues. We focus on a case study for medical entities motivated by our findings in previous work. In [1] we investigated the semantic properties of Word2Vec for pharmaceutical entity-relations and subsequently utilized them as an alternative access path for the pharmaceutical digital library PubPharm[1]. In brief, we found that semantically meaningful drug-to-drug relations are indeed reflected in the high-dimensional word embeddings. Here, we aim to identify the effect of scaling methods such as Multidimensional Scaling (MDS) and Principal Component Analysis (PCA) on active substance embeddings learned by Word2Vec.

In the following, we show that scaling has a high correlation with the number of Word2Vec training dimensions. This finding means that by using scaling, we can find where the optimal number of training dimensions regarding purity, precision, and recall is located. Our results can be of interest for all approaches in which Word-Embedding

---

[1] https://www.pubpharm.de/vufind/.

training has to be applied to massive amounts of data (Big Data) and thus exploring different numbers of dimensions with re-training is not a practical option.

The paper is organized as follows: Sect. 2 revisits related work accompanied by our extensive investigation of scaling approaches to embedded drug-clusters in Sect. 3. We close with conclusions in Sect. 4.

2 Related Work

Neural Language Model Representation of Words. Semantic embeddings of words in vector spaces have sparked interest, especially Word2Vec [11, 12] and related methods [13–16]. Researchers have demonstrated that words with similar meanings are embedded nearby, and even 'word arithmetic' can be convincingly applied. For example, the calculated difference in the embedding vector space between 'Berlin' and 'Germany' is similar to the one obtained between 'Paris' and 'France'. Word2Vec representations are learned in an unsupervised manner from large corpora and are not explicitly constrained to abide by such regularities. In a nutshell, Word2Vec is a technique for building a neural network that maps words to real number vectors. What is unique about these number vectors is that words with similar meaning will map to similar vectors. At its core, Word2Vec constructs a log-linear classification network. More specifically, in [11, 12] researchers proposed two such networks: the Skip-gram and the Continuous Bag-of-Words (CBoW). In our experiments we used the Skip-gram architecture, which is considered preferable according to the experiments reported by [12].

Multidimensional Scaling (MDS). Multidimensional Scaling [17] deals with the problem of representing a set of n objects in a low-dimensional space in which the distances respect the distances in the original high-dimensional space. In its classical formalization MDS takes as input a dissimilarity matrix between pairs of objects and outputs a coordinate matrix whose configuration minimizes a loss function called stress or strain [17]. In our experimental setting, given a matrix of the Euclidean distances between entities represented by Word2Vec vectors, $M = [ed_{i,j}]$ where $ed_{i,j}$ is the distance between the pair of entities i, j. MDS uses eigenvalue decomposition on the matrix M using double centering [18]. In our experiments we used the Scikit-Learn Python implementation [19] with default parameters except for the number of dimensions that we exhaustively tested.

Principal Component Analysis (PCA). Principal Component Analysis is a popular data mining technique for dimensionality reduction [25]. Given a set of data points on n dimensions, PCA aims to find a linear subspace of dimension d lower than n such that the data points lie mainly on this linear subspace. In our case we take the matrix M_e of Word2Vec vectors where the rows represent medical entities and columns to the dimensions of the Word2Vec semantic space. The idea of PCA then is to treat the set of tuples in this matrix and find the eigenvectors for $M_e M_e^T$. When you apply this transformation to the original data, the axis corresponding to the principal eigenvector is the one along which the points are most spread out. In other words, this axis is the one along which the variance of the data is maximized. Thus, the original data is approximated by data that has many fewer dimensions and that summarizes well the original data.

Orthogonal Procrustes. We use Orthogonal Procrustes [23] – also known as rotational alignment – to evaluate the relative quality of two different scaling approaches. The general idea here is to evaluate two scaling techniques without considering any specific metric related to the clustering task. Instead it is assessed by measuring pointwise differences, which of the two scaling approaches can better approximate the original Word2Vec space. Orthogonal Procrustes was used before to align word embeddings created at different time periods, i.e., to analyze semantic changes of words in diachronic embeddings [21].

3   Investigation of Effect of the Dimensionality Reduction

First, we describe the methodology for generating our ground truth dataset. After this, we describe our ground truth evaluation corpus followed by experimental set-up and implementation decisions. Then we examine with the help of our ground truth dataset whether the number of Word2Vec training dimensions and the number of scaling dimensions correlate with purity, precision, recall, and F-Score. We will then perform a mathematical analysis between MDS, PCA, and Word2Vec results based on statistical t-test and matrix approximation methods. Afterwards we compare the runtime of MDS, PCA and the training with different Word2Vec dimensions. Since our current study is based on the dataset of our previous work [1], we use almost the same methodology, evaluation corpus, implementation, and set-up decisions:

Methodology for Building our Ground-Truth Dataset
After the initial crawling step the following process can be roughly divided into four sub-steps:

1. *Preprocessing of crawled documents.* After the relevant documents were crawled, classical IR-style text pre-processing is needed, i.e., stop-word removal and stemming. The pre-processing helps mainly to reduce vocabulary size, which leads to improved performance, as well as improved accuracy. Due to their low discriminating power, all words occurring in more than 50% of the documents are removed. Primarily, these are often used words in general texts such as '*the*' or '*and*', as well as terms frequently used within a domain (as expressed by the document base), e.g., '*experiment*', '*molecule*', or '*cell*' in biology. Stemming further reduces the vocabulary size by unifying all flections of terms. A variety of stemmers for different applications is readily available.
2. *Creating word embeddings for entity contextualization.* Currently, word embeddings [10] are the state-of-the-art neural language model technique to map terms into a multi-dimensional space (usually about 200-400 dimensions are created), such that terms sharing the same context are grouped more closely. According to the distributional hypothesis, terms which often share the same context in larger samples of language data, in general also share similar semantics (i.e., have a similar meaning). In this sense, word embeddings group entities sharing the same

context and thus collecting the nearest embeddings of some search entity leads to a group of entities sharing similar semantics.

3. *Filtering according to entity types.* The computed word embeddings comprise at this point a significant portion of the corpus vocabulary. For each vocabulary term there is precisely one-word vector representation as the output of the previous step. Each vector representation starts with the term followed by individual values for each dimension. In contrast, classical facets only display information of the same type, such as publication venues, (co-)authors, or related entities like genes or enzymes. Thus, for the actual building of facets, we need only vector representations of the same entity type. Here, dictionaries are needed to sort through the vocabulary for each type of entity separately. The dictionaries either can be directly gained from domain ontologies, like, e.g., MeSH for illnesses, can be identified by named entity recognizers, like e.g., the Open Source Chemistry Analysis Routines (OSCAR, see [24]) for chemical entities, or can be extracted from open collections in the domain, like the DrugBank for drugs.

4. *Clustering entity vector representations.* The last step is preparing the actual grouping of entities closely related to each other. To do this, we apply a k-means clustering technique on all embedded drug representations and decide for optimal cluster sizes: in our approach optimal cluster sizes are decided according to the *Anatomical Therapeutic Chemical (ATC) Classification System*[2]. Here ATC subdivides drugs according to their anatomical properties, therapeutic uses, and chemical features.

Experimental Ground-Truth Dataset Setup

Evaluation corpus. With more than 27 million document citations, PubMed[3] is the largest and most comprehensive digital library in the bio-medical field. However, since many documents citations do not feature full texts, we relied solely on abstracts for learning purposes. As an intuition, the number of abstracts matching each pharmaceutical entity under consideration should be 'high enough' because with more training data, more accurate contexts can be learned, yet the computational complexity grows. Thus, we decided to use the 1000 most relevant abstracts for each entity according to the relevance weighting of PubMed's search engine [29].

Query Entities. As query entities for the evaluation, we randomly selected 275 drugs[4] from the *DrugBank*[5] collection. We ensured that each selected drug featured at least one class label in ATC and occurred in at least 1000 abstracts on PubMed. Thus, our final document set for evaluation contained 275,000 abstracts. Therefore, these drugs usually have a *one-word* name, which makes it straightforward to filter them out after a Word2Vec training iteration. However, besides our specific case, pharmaceutical entities often consist of several words (e.g., diabetes mellitus) and can also have many synonyms (e.g., aspirin/acetylsalicylic acid). Phrases and synonyms are a general

---

[2] https://www.whocc.no/atc_ddd_index/.

[3] https://www.ncbi.nlm.nih.gov/pubmed/.

[4] The complete list can be downloaded under: http://www.ifis.cs.tu-bs.de/webfm_send/2295.

[5] https://www.drugbank.ca/.

problem for word embedding algorithms because they are usually trained on single words, resulting in one vector per word and not per entity. A possible solution for such cases is (1) applying named entity recognition in documents and (2) placing a unique identifier at the entity's position in the text. Here, entity recognition can be done using PubTator[6], which is a tool that can recognize pharmaceutical entities as well as their position in text and that returns a unique MeSH-Id for each of them.

As ground truth, all class labels were crawled from *DrugBank*. Since the ATC classification system shows a fine-grained hierarchical structure, we remove all finer levels before assigning the respective class label to each drug. For example, one of the ATC classes for the drug 'Acyclovir' is '*D06BB53*'. The first letter indicates the main anatomical group, where '*D*' stands for 'dermatological'. The next level consists of two digits '*06*' expressing the therapeutic subgroup 'antibiotics and chemotherapeutics for dermatological use'. Each further level classifies the object even more precisely, until the finest level usually uniquely identifies a drug. In our active ingredient collection there are 13 different ATC class labels of the highest level. We use these 13 different labels to divide the 275 active ingredients into 13 (*ground truth*) clusters.

Ground Truth Dataset Implementation and Parameter Settings

1. *Text Preprocessing:* Stemming and stop-word removal were performed using a *Lucene*[7] index. For stemming we used Lucene's *Porter Stemmer* implementation.
2. *Word Embeddings*: After preprocessing, word embeddings were created with Gensims's *Word2Vec*[8] implementation. To train the neural network, we used a minimum word frequency of 5 occurrences. We set the word window size to 20 and the initial layer size to 275 features per word. Training iterations were set to 4.
3. *Entity filtering*. While Word2Vec generated a comprehensive list of word vector representations, we subsequently filtered out all vectors not related to any Drug-Bank entity (resulting in 275 entity-vectors).
4. *Clustering vector representations*. In this step we clustered the 275 entity vector representations obtained in the previous filtering step in 13 clusters. For the clustering step we used Python [19] Multi-KMean ++ implementation.

3.1   Experimental Investigation

First, we need to clarify how a correlation between the different approaches can be measured. We also need to determine whether the scaling approaches are faster. In this context, the following quality criteria should be fulfilled:

- *Empirical Correlation Accuracy*: The result of a scaling approach should be comparable to the result of a Word2Vec training for a fixed number of training dimensions. Therefore, we will always determine the 'semantic quality' of a semantic space by evaluating purity, F-Score, precision, and recall against the ground truth expressed by the ATC classification. After scaling down the original

---

[6] https://www.ncbi.nlm.nih.gov/CBBresearch/Lu/Demo/PubTator/.

[7] https://lucene.apache.org/.

[8] https://radimrehurek.com/gensim/models/word2vec.html.

Word2Vec space trained on 275 dimensions to n dimensions (where $n < 275$) the semantic quality of this space needs to be compared to the respective quality of a Word2Vec space directly trained using only n dimensions. Are the respective qualities correlated for different values of n?

- *Mathematical Accuracy:* The result of a scaling should resemble the vectors of a Word2Vec training. A similarity between the vectors would underpin the results of our empirical study as well as help us to find possible differences between PCA and MDS. To test our hypothesis, we perform a mathematical analysis based on statistical t-test and matrix approximation using orthogonal Procrustes.
- *Scaling Performance*: Performing a scaling iteration for some number of dimensions should on average be significantly faster than training a Word2Vec model using the same number of dimensions.

3.2 Empirical Correlation Accuracy

In our first experiment we investigate if scaling with MDS and PCA correlates with the number of Word2Vec training dimensions regarding the following quality measures: F-Score, precision, recall, and purity. We determine the quality measures for our clusters using the method described in Manning et al. [20]. Initially we train Word2Vec using 275 dimensions, and we choose the maximum of 275 dimensions because of the technical implication for calculating PCA. Technically speaking there exist a Principal Component for each variable in the data set. However, if there are fewer samples than variables, then the number of samples puts an upper bound on the number of Principal Components with eigenvalues greater than zero [25]. Therefore, for this experiment, we perform the following steps for each number of dimensions n (where $n < 275$):

- *Scaling Step*: First we scale the initially trained and filtered 275 active substance Word2Vec vectors with dimensions n using MDS and PCA. Also, we train *Word2Vec* with n dimensions on the evaluation corpus and then filter out the 275 active substance vectors.
- *Clustering Step*: For each of the three results from the previous step, we assign each active ingredient to one of the possible 13 ATC class labels. Then we perform clustering with $k = 13$ and a total of 50 iterations. In each clustering iteration we calculate the quality measures mentioned above and calculate the mean values for purity, precision, recall, and F-Score.

Figure 1 shows the respective mean values regarding each quality measure for the different choices of dimensions. Table 1 lists the correlation (Pearson correlation coefficient) values between the different methods. The mean values of the individual dimensions were used for the correlation calculation. As can be seen, there is a strong correlation for all values, whereby the values for MDS correlate best with the Word2Vec result. Thus, scaling approaches indeed lead to similar results as Word2Vec training.

Can the optimal training dimension be determined using a scaling method? As can be seen, the highest mean values (Fig. 1) of the different methods are almost precisely in the same dimension range (e.g., precision). This observation allows us to predict the optimum number of training dimensions quite accurately using scaling approaches. *Is*

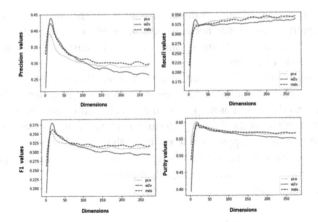

Fig. 1. Precision, recall, F1, and purity mean values for PCA, MDS, and Word2Vec.

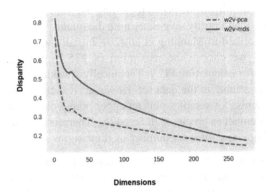

Fig. 2. Disparity comparison (lower values are better) between Word2Vec-PCA and Word2Vec-MDS using procrustes analysis.

Table 1. Correlation (Pearson correlation coefficient) values between the different approaches. Where PCC is the correlation coefficient between precision values, RCC between recall values, F1CC between F1-Values, and PuCC is the correlation coefficient between purity values.

Correlation between	PCC	RCC	F1CC	PuCC
MDS-W2V	0.90	0.80	0.85	0.87
PCA-W2V	0.85	0.78	0.81	0.69

the quality comparable? Surprisingly, a Word2Vec training does not always lead to the best result. For example, we can observe that scaling for most dimensions (~ 200) leads to a better result. In particular, we achieve the best purity-value with PCA. In short, it probably pays off to use a scaling approach. The differences for the other quality measures are rather small. For example, MDS can only achieve a $\sim 2\%$ worse precision result, but on the other hand, MDS scaling alone can increase the precision values by up to ~60%, and F1-Values up to 20%. What we can also see is that our optimum for all quality measures lies at about 25 dimensions. This value, in turn, deviates quite far from the recommended 200–400 dimensions for a Word2Vec training. Our finding indicates that for a particular problem domain, as in our case, a standard choice of dimensions for a Word2Vec training can be a disadvantage.

3.3   Mathematical Accuracy

We performed two evaluations to assess the quality of the scaling approaches that we compared with Word2Vec. The first evaluation corresponds to what we called metric-based analysis because specific metrics that depend on the task at hands such as precision, recall, and F1 are needed. In contrast, non-metric based evaluation considers only the approximation quality of the scaling algorithms regarding the original Word2Vec space.

Metric-Based Analysis. In this first evaluation we used precision, recall, and F-Score to perform a pair-wise t-test comparison. With a 95% confidence interval the differences between PCA and MDS are not statistically significant for precision and F-Score. However, in recall, the differences between PCA and MDS are statistically significant.

To provide the reader with a visual interpretation of the results found in the recall t-test, we show in Figs. 3, 4, and 5 the Bland-Altman Plots [22] that compares MDS with PCA, MDS with Word2Vec, and PCA with Word2Vec, respectively. Bland-Altman plots compare in a simple plot two measurements to ascertain if indeed differences exist between them. In the x-axis the graph shows the mean of the measurements and on the y-axis their differences. Thus, if there are no differences, we should observe on the y-axis that most of the values are near zero. This type of plot makes it easier to see if there are magnitude effects, for example when small values of x and y are more similar than large values of x and y. We can observe in that differences between PCA and Word2Vec are negligible regarding recall values. Moreover, we can observe that the higher the values of recall, the better PCA is in approximating Word2Vec. In summary, the plot (Fig. 5) shows that PCA leads to a slightly better approximation of recall values than MDS.

Non-Metric Based Analysis. Finally to evaluate the differences between MDS and PCA, we decided to assess the approximation power of the two methods using Procrustes analysis. This analysis complements our previous metric-based analysis by introducing an evaluation of the MDS and PCA spaces regarding how good each of them can approximate the original Word2Vec space. What we mean here by Procrustes analysis is the following: given two identical sized matrices, Procrustes tries to apply transformations (scaling, rotation and reflection) on the second matrix to minimize the sum of squares of the pointwise differences between the two matrices (disparity hereafter). Put another way, Procrustes tries to find out what transformation of the second matrix can better approximate the first matrix. The output of the algorithm is not only the transformation of the second matrix that best approximates matrix one but also the disparity between them.

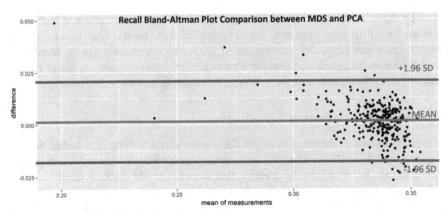

Fig. 3. Bland-Altman plot using recall measures PCA vs MDS.

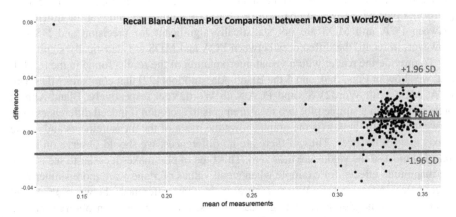

Fig. 4. Bland-Altman plot using recall measures MDS vsWord2Vec.

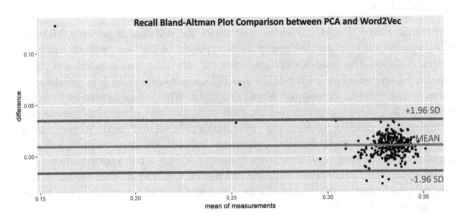

Fig. 5. Bland-Altman plot using recall measures PCA vs Word2Vec.

We use the disparity value in our analysis to determine which of the scaling algorithms can better approximate Word2Vec original space. Low disparity values are better by definition. In Fig. 2 we plot the disparity values using dimensions up to 275 which is the maximum number that we can use because we have only 275 active substances as our input matrix. To generate the plot, we train Word2Vec for dimensions two up to 275. We used the original space of 275 dimensions from Word2Vec to apply MDS and PCA using dimension from 2 up to 275. Thus, each point in the plot shows the disparity value between the corresponding scaling algorithm and Word2Vec. We can see that PCA outperforms MDS because it shows lower disparity values for each of the dimensions calculated. In other words, PCA preserves the quality of the semantics of the original Word2Vec space better than MDS.

3.4   Scaling Performance

After having shown that there is both a strong empirical as well as a robust mathematical correlation between scaling approaches and a Word2Vec training using the same number of dimensions, we then compare the runtime performance of the different approaches. Here, we first train Word2Vec on our ground truth corpus with 275 dimensions and extract the 275 active substances vectors again. Then we scale the result with PCA and MDS to dimensions n (where $n < 275$), after which we measure the cumulative time which was required for scaling to all number of dimensions. Also, we train Word2Vec with the different number of dimensions n ($n < 275$) and measure the cumulative training time for comparison with the scaling approaches. This kind of Word2Vec training corresponds to the usual procedure to determine an optimal result (e.g., regarding F-Score). All three calculations are performed one after the other on the same computer with the following characteristics: 16 Xeon E5/Core i7 Processors with 377 GB of RAM. The results of our experiments are shown in Table 2:

Table 2. Runtime *(seconds)*: Sum of the runtimes of the different approaches in seconds. *Runtime reduction*: Reduction in % of run times compared to a Word2Vec (W2V) training

Approach	Runtime (seconds)	Runtime reduction
PCA	17	99.83%
MDS	1162	88.22%
W2V	9865	-

As can be seen in Table 2, scaling approaches need significantly less time on our active substance dataset. Here, a runtime reduction of up to 99% can be achieved. PCA was much faster in scaling compared to MDS. Given the observed runtime reduction, it pays off to use scaling approaches when training on a large corpus.

4 Conclusions

We have conducted an experimental analysis of scaling algorithms applied over a set of entities using neural language models for clustering purposes. Indeed, one of the most critical parameters of implementations such as Word2 Vec is the number of training dimensions for the neural network. Because different testing numbers are time-consuming and thus can take hours or even days per training iteration on large text corpora, we have investigated an alternative using scaling approaches. In particular, we used the implementation provided by Word2 Vec and contrasted Multidimensional Scaling and Principal Component Analysis quality. We conclude here by summarizing our main findings for researchers and practitioners looking to use Word2 Vec in similar problems.

Our experiments indicate that there exists a strong correlation (up to 90%) regarding purity, F1, as well as precision and recall. We have shown that for a particular problem domain, as in our active substance case, a standard choice of dimensions for a Word2 Vec training can be a disadvantage. Moreover, by mathematical analysis we have shown that the spaces after scaling strongly resemble the original Word2Vec semantic spaces. Indeed, the quality of the scaling approaches is quite comparable to the original Word2Vec space: they achieve almost the same precision, recall, and F1 measures.

As a performance bonus, we have shown that performance of scaling approaches regarding execution times is several orders of magnitude superior to Word2Vec training. For instance, we obtained more than 99% of time-saving when computing PCA instead of Word2Vec training. Researchers could thus rely on initial Word2Vec training or pre-trained (Big Data) models such as those available for the PubMed[9] corpus or Google News[10] with high numbers of dimensions and afterward apply scaling approaches to quickly find the optimal number of dimensions for any task at hand.

References

1. Wawrzinek, J., Balke, W.-T.: Semantic facettation in pharmaceutical collections using deep learning for active substance contextualization. In: Choemprayong, S., Crestani, F., Cunningham, S.J. (eds.) ICADL 2017. LNCS, vol. 10647, pp. 41–53. Springer, Cham (2017). https://doi.org/10.1007/978-3-319-70232-2_4
2. Wang, Z.Y., Zhang, H.Y.: Rational drug repositioning by medical genetics. Nat. Biotechnol. **31**(12), 1080 (2013)
3. Abdelaziz, I., Fokoue, A., Hassanzadeh, O., Zhang, P., Sadoghi, M.: Large-scale structural and textual similarity-based mining of knowledge graph to predict drug–drug interactions. Web Semant.: Sci., Serv. Agents World Wide Web **44**, 104–117 (2017)
4. Leser, U., Hakenberg, J.: What makes a gene name? Named entity recognition in the biomedical literature. Brief. Bioinform. **6**(4), 357–369 (2005)

---

[9] https://github.com/RaRe-Technologies/gensim-data/issues/28.

[10] https://code.google.com/archive/p/word2vec/.

5. Lotfi Shahreza, M., Ghadiri, N., Mousavi, S.R., Varshosaz, J., Green, J.R.: A review of network-based approaches to drug repositioning. Brief. Bioinform. bbx017 (2017)
6. Dudley, J.T., Deshpande, T., Butte, A.J.: Exploiting drug–disease relationships for computational drug repositioning. Brief. Bioinform. **12**(4), 303–311 (2011)
7. Willett, P., Barnard, J.M., Downs, G.M.: Chemical similarity searching. J. Chem. Inf. Comput. Sci. **38**(6), 983–996 (1998)
8. Ngo, D.L., et al.: Application of word embedding to drug repositioning. J. Biomed. Sci. Eng. **9**(01), 7 (2016)
9. Lengerich, B.J., Maas, A.L., Potts, C.: Retrofitting Distributional Embeddings to Knowledge Graphs with Functional Relations. arXiv preprint arXiv:1708.00112 (2017)
10. Baroni, M., Dinu, G., Kruszewski, G.: Don't count, predict! a systematic comparison of context-counting vs. context-predicting semantic vectors. In: Proceedings of the 52nd Annual Meeting of the Association for Computational Linguistics, Long Papers, vol. 1, pp. 238–247 (2014)
11. Mikolov, T., Yih, W.T., Zweig, G.: Linguistic regularities in continuous space word representations. In: Proceedings of the 2013 Conference of the North American Chapter of the Association for Computational Linguistics: Human Language Technologies, pp. 746–751 (2013)
12. Mikolov, T., Chen, K., Corrado, G., Dean, J.: Efficient estimation of word representations in vector space. In: NIPS (2013)
13. Levy, O., Goldberg, Y.: Neural word embedding as implicit matrix factorization. In: Advances in Neural Information Processing, pp. 2177–2185 (2014)
14. Pennington, J., Socher, R., Manning, C.: Glove: global vectors for word representation. In: Proceedings of the 2014 Conference on Empirical Methods in Natural Language Processing (EMNLP), pp. 1532–1543 (2014)
15. Bengio, Y., Courville, A., Vincent, P., Collobert, R., Weston, J., et al.: Natural language processing (almost) from scratch. IEEE Trans. Pattern Anal. Mach. Intell. **35**, 384–394 (2014)
16. Joulin, A., Grave, E., Bojanowski, P., Mikolov, T.: Bag of tricks for efficient text classification, vol. 2, pp. 427–431 (2016). Proceedings of the 15th Conference of the European Chapter of the Association for Computational Linguistics. Valencia, Spain, 3–7 April 2017
17. Borg, I., Groenen, P.J.: Modern Multidimensional Scaling: Theory and Applications. Springer, New york (2005). https://doi.org/10.1007/0-387-28981-X
18. Weinberg, S.L.: An introduction to multidimensional scaling. Meas. Eval. Couns. Dev. **24**, 12–36 (1991)
19. Pedregosa, F., Varoquaux, G., Gramfort, A., Michel, V., et al.: Scikit-learn: machine learning in Python. J. Mach. Learn. Res. **12**, 2825–2830 (2011)
20. Manning, C.D., Raghavan, P., Schütze, H.: Introduction to Information Retrieval. Cambridge University Press, Cambridge (2008)
21. Hamilton, W.L., Leskovec, J., Jurafsky, D.: Diachronic word embeddings reveal statistical laws of semantic change. In: Proceedings of the 54th Annual Meeting of the Association for Computational Linguistics, Berlin, Germany, 7–12 August 2016, pp. 1489–1501 (2016)
22. Altman, D.G., Bland, J.M.: Measurement in medicine: the analysis of method comparison studies. Statistician **32**, 307–317 (1983)
23. Schönemann, P.H.: A generalized solution of the orthogonal procrustes problem. Psychometrika **31**, 1–10 (1966)
24. Jessop, D.M., Adams, S.E., Willighagen, E.L., Hawizy, L., Murray-Rust, P.: OSCAR4: a flexible architecture for chemical text-mining. J. Cheminformatics **3**(1), 41 (2011)

25. Leskovec, J., Rajaraman, A., Ullman, J.D.: Mining of Massive Datasets. Cambridge University Press, Cambridge (2014)
26. Levy, O., Goldberg, Y.: Neural word embedding as implicit matrix factorization. In: Advances in Neural Information Processing Systems, pp. 2177–2185 (2014)
27. Gittens, A., Achlioptas, D., Mahoney, M.W.: Skip-gram - zipf + uniform = vector additivity. In: Proceedings of the 55th Annual Meeting of the Association for Computational Linguistics, Long Papers, vol. 1, pp. 69–76 (2017)
28. Li, Y., Xu, L., Tian, F., Jiang, L., Zhong, X., Chen, E.: Word embedding revisited: a new representation learning and explicit matrix factorization perspective. In: IJCAI International Joint Conference on Artificial Intelligence, pp. 3650–3656 (2015)
29. Canese, K.: PubMed relevance sort. NLM Tech. Bull **394**, e2 (2013)

Combining Semantic and Lexical Measures to Evaluate Medical Terms Similarity

Silvio Domingos Cardoso[1,2(✉)], Marcos Da Silveira[1], Ying-Chi Lin[3],
Victor Christen[3], Erhard Rahm[3], Chantal Reynaud-Delaître[2],
and Cédric Pruski[1]

[1] LIST, Luxembourg Institute of Science and Technology,
Esch-sur-Alzette, Luxembourg
{silvio.cardoso,marcos.dasilveira,cedric.pruski}@list.lu
[2] LRI, University of Paris-Sud XI, Gif-sur-Yvette, France
chantal.reynaud@lri.fr
[3] Department of Computer Science, Universität Leipzig, Leipzig, Germany
{lin,christen,rahm}@informatik.uni-leipzig.de

Abstract. The use of similarity measures in various domains is corner-
stone for different tasks ranging from ontology alignment to information
retrieval. To this end, existing metrics can be classified into several categories among which lexical and semantic families of similarity measures
predominate but have rarely been combined to complete the aforemen-
tioned tasks. In this paper, we propose an original approach combining
lexical and ontology-based semantic similarity measures to improve the
evaluation of terms relatedness. We validate our approach through a set
of experiments based on a corpus of reference constructed by domain
experts of the medical field and further evaluate the impact of ontology
evolution on the used semantic similarity measures.

Keywords: Similarity measures · Ontology evolution
Semantic Web · Medical terminologies

1 Introduction

Measuring the similarity between terms is at the heart of many research investigations. In ontology matching, similarity measures are used to evaluate the
relatedness between concepts from different ontologies [9]. The outcomes are the
mappings between the ontologies, increasing the coverage of domain knowledge
and optimize semantic interoperability between information systems. In information retrieval, similarity measures are used to evaluate the relatedness between
units of language (e.g., words, sentences, documents) to optimize search [39].

This work is supported by FNR Luxembourg and DFG Germany through the ELISA
project.

© Springer Nature Switzerland AG 2019
S. Auer and M.-E. Vidal (Eds.): DILS 2018, LNBI 11371, pp. 17–32, 2019.
https://doi.org/10.1007/978-3-030-06016-9_2

The literature of this domain reveals that several families of similarity measures can be distinguished [14,16] such as string-based, corpus-based, knowledge-based metrics, etc. Lexical Similarity Measures (LSM) regroups the similarity families that rely on syntactic or lexical aspects of the units of languages [29]. Such metrics are efficient to compare strings such as *"Failure of the kidney"* with *"Kidney failure"*. However, they do not capture very well the semantic similarity. For instance, *"Cancer"* and *"malignancy"* can be totally disjointed from the lexical point of view despite their closely related semantics. To overcome this barrier, Semantic Similarity Measures (SSM) have been introduced. They exploit meaning of terms to evaluate their similarity. This is done using two broad types of semantic proxies: corpora of texts and ontologies.

The corpora proxy uses Information Content (IC-based) to observe the usage of terms and determine the similarity based on the distribution of the words or the co-occurrence of terms [25,26]. The ontology proxy uses the *intrinsic* Information Content (*i*IC-based) [16], where the structure of the ontology allows calculating some semantic similarities [6]. Both proxies have been use in several domains, but we are working mainly with ontologies and we focus our analysis on SSM that are *i*IC-based. Although single similarity measures have been successfully used in many works, their combination remains under explored, especially the couple LSM/*i*IC-based SSM. The goal of this work is not to propose another similarity measure, but to demonstrate that weighted combination of existing ones can improve the outcomes.

Our motivation for this work came from the observations, in our previous work on semantic annotations and mappings adaptation [2,3,11], that few information is made available to understand how mappings and semantic annotations were generated and how they are maintained over time. In order to propose an automatic maintenance method for mappings and annotations [4], we search for patterns that allow reasonable explanations for the selection of terms and their relations. The similarity metrics became an essential tool for our approach. We deal with datasets of mapping and annotations that were generated based on very different methods (automatically and/or manually). We are interested on finding a combination of methods that can better explain the reasoning behind the generation/maintenance of mappings or annotations. The single method (LSM or SSM) that we evaluated did not represent well the patterns that we are looking for. Thus, we empirically evaluated the SSM×LSM combination and we are presenting the outcomes of our research in this paper. Differently from other comparative approaches that look for unifying methods or automatically select the best single method for a specific task, the goal of our research was to look for combinations of methods. Our ultimate goal is to define a procedure to analyze the evolution of ontologies and collect relevant information to be used to preserve the validity of annotations and mappings.

In this paper, we present a weighting method that combines LSM and ontology-based SSM to evaluate the relatedness between terms (word or multi-token terms) in order to improve the characterization of changes occurring in the ontology at evolution time. We based our solution on existing well known similarity measures and evaluate it using a Gold standard corpus. Our iterative method allows to find the best weights to associate to LSM and SSM respectively. In our experiments we used datasets constructed by experts from the medical domain [32]. It gathers scores given by domain experts on the relatedness between terms. We proved the validity of our measure by first showing the correlation between the obtained values and the scores given by domain experts on the data of the reference corpus using the Spearman's rank correlation metric. We then use the Fisher's Z-Transformation to evaluate the added value of our metric with respect to state-of-the-art similarity measures. Through this work, we are able to show:

- The added value of combining LSM and ontology-based SSM for measuring term relatedness.
- The validity of the combination SSM×LSM with respect to experts score.
- The impact of the evolution of ontologies on the used SSM.
- The most suitable metrics and weights for SNOMED CT and MeSH.

The remainder of this article is structured as follows. Section 2 introduces the various concepts needed to understand our approach. This includes the definition of existing lexical and semantic similarity measures as well as the methods we have followed to evaluate our work. Section 3 presents the related work. Section 4 describes our approach for combining lexical and semantic similarity measures as well as our evaluation methodology while results are presented in Sect. 5. Section 6 discuss the results. Finally, Sect. 7 wraps up with concluding remarks and outlines future work.

2   Background

In this section, we provide the necessary background information to understand the notion tackled in this paper. We start by listing the studied LSM and SSM. We then explain the Spearman's rank correlation and the Fisher's Z-transformation formulas we have used in our experiments.

2.1   Lexical Similarity Measures

In our work, we introduced lexical similarity measures through various string-based approaches. It consists in the analysis of the composition of two strings to determine their similarity. Two types of approach can be distinguished: character-based and term-based. The former denotes the comparison of two strings and the quantification of the identified differences. The latter compares the differences between words composing the string. In our experiments, we have used the 12 following LSM: Levenshtein, Smith-Waterman, Jaccard, Cosine, Block Distance, Euclidean Distance, Longest Common Substring, Jaro-Winkler, LACP, TF/IDF, AnnoMap [5] and Bigram.

2.2 Semantic Similarity Measures

Semantic similarity measures denote a family of metrics that rely on external knowledge to evaluate the distance between terms from their meaning point of view. It encompasses corpus-based metrics and ontology-based which [15]. In this work, we put the stress on ontology-based approaches (iIC-based). We have retained 11 SSMs following a deep literature survey. Table 1 hereafter contains the various semantic similarity measures that have been tested in our work. Table 2 refers to iIC-based methods. Note that the SSMs methods from Table 1 use as input the outcomes of iIC-based methods. Thus, when presenting the results we indicate the name of the SSM method as well as the iIC-based method used as input.

Table 1. Used semantic similarity measures

SSM	Description
Jiang Conrath [19]	Similar to Resnik, it uses a corpus of documents in addition to an ontology
Feature Tversky Ratio Model [40]	Considers the features of label to compute similarity between different concepts, but the position of the concept in the ontology is ignored. Common features tend to increase the similarity and other features tend to decrease the similarity
Tversky iIC Ratio Model [8]	
Lin [21]	Similar to Resnik's measure but uses a ratio instead of a difference
Lin GraSM [7]	
Mazandu [23]	Combination of node and edge properties of Gene Ontology terms
Jaccard iIC [17]	It consists in the ratio between the intersection of two sets of feature and the union of the same two sets
Jaccard 3W iIC [27]	
Resnik GraSM [35]	See Table 2
Resnik [35]	
Sim iIC [20]	Exploits iIC of the Most Informative Common Ancestor of the concepts to evaluate

2.3 Spearman's Rank Correlation

One of the objectives of this work is to experimentally show the complementarity of LSM and ontology-based SSM to better evaluate the relatedness between terms. Since we compared the results obtained experimentally with the score assigned by domain experts, we need a method to evaluate their correlation. Spearman's Rank Correlation (cf. Eq. 1) is a statistical method that measures the coefficient strength of a linear relationship between paired data [34]. In other words, its verifies whether the values produced by automatic similarity measures and scores given by domain specialists are correlated.

Table 2. Information Content based measures

iIC-based metrics	Description
Resnik (normalized) [12]	Based on the lowest common ancestor
Sanchez [36]	iIC of a concept is directly proportional to its number of taxonomical subsumers and inversely proportional to
Sanchez adapted [36]	the amount of leaves of its hyponym tree
Seco [38]	iIC is computed based on the number of hyponyms a concept has in WordNet. This metric does not rely on corpus
Zhou [41]	iIC considers not only the hyponyms of each word sense in WordNet but also its depth in the hierarchy
Harispe [17]	Modification of [36] in order to authorize various non uniformity of iICs among the leafs
Max depth non linear [17]	iIC of a concept is directly computed based on the
Max depth linear [17]	depth of the concept
Ancestors Norm [17]	iIC of a concept is computed based on the number of ancestors of the concept

$$r_s = 1 - \frac{6 \sum_i d_i}{n(n^2 - 1)} \tag{1}$$

In Eq. 1, d_i is the difference between the two ranks of each observation and n is the number of observations.

2.4 Fisher's Z-Transformation

Fisher's Z-Transformation is a statistic method that allows us to verify whether two nonzero's Spearman's rank coefficients are statistically different [34]. The corresponding formula is:

$$z = \frac{1}{2} \ln \left(\frac{1 + r_s}{1 - r_s} \right) \tag{2}$$

Through this normalization of Spearman's rank coefficient we can assure whether r_s from an automatic similarity method X_i is better than a r'_s from a method Y_i.

In order to compare various correlations, we have to apply the following three-steps method:

– Conversion of r_s and r'_s to z_1 and z_2 by applying Eq. 2.
– Compute the probability value $\rho \in 0 \le \rho \le 1$ through Eq. 3, where N_1 and N_2 are the number of elements in our dataset and $erfc$ denotes the complementary error function.

- Test the null assumption $H_0 : r_s = r'_s$ case $\rho > 0.05$ and vice versa. Nevertheless, it only can be performed when N, i.e., the number of paired data is moderately large ($N \geq 10$) to assure the statistical significance.

$$\rho = erfc \left(\frac{|z_1 - z_2|}{\sqrt{2}\sqrt{\frac{1}{N_1-3} + \frac{1}{N_2-3}}} \right) \tag{3}$$

In consequence, a value smaller than 0.05 indicates that the two evaluated measures are statistically different.

3   Related Work

SSM and LSM have been widely used in order to evaluate the relatedness between terms specially in the biomedical domain. However, the combination of LSM and ontology-based SSM has rarely been investigated. Relevant initiatives were proposed in [22,32] where authors have adapted WordNet based similarity measures to the biomedical domain. Lord et al. [22] have focused on Gene Ontology, a domain specific ontology, while Pedersen et al. [32] decided to be more generic and have grounded their work on SNOMED CT and reinforce their metrics with information derived from text corpora.

In the same line, the authors of [18] present a method for measuring the semantic similarity of texts combining a corpus-based measure of semantic word similarity and a normalized and modified version of the longest common subsequence string matching algorithm. They further evaluate the proposed metric on two well-accepted general corpus of text and show the added value of the combination with respect to comparable existing similarity measures.

In [17], the authors have investigated a broad range of semantic similarity measures to identify the core elements of the existing metrics with a particular focus on ontology-based measures. They further came up with a framework aiming at unifying the studied metrics and show the usability of the framework on the same corpus that is used in the work of Petersen et al. [32].

Aouicha and Taieb [1] exploit the structure of an ontology to achieve a better semantic understanding of a concept. Their Information Content-based semantic similarity measure consists in expressing the IC by weighting each concept pertaining to the ancestors' subgraph modeling the semantics of a biomedical concept. They validated the added value of their work on three datasets including the one we are using in this work [32].

The work presented in [37] classifies ontology-based semantic similarity measures. They distinguish between edge-counting approaches, Feature-based approaches and intrinsic content ones. Moreover, they defined another ontology-based measure. Their metric considers as features the hierarchy of concepts structuring the ontology in order to evaluate the amount of dissimilarity between concepts. In other words, they assume that a term can be semantically different from other ones by comparing the set of concepts that subsume it.

Oliva et al. [29] have defined the SyMSS method consisting in assessing the influence of the syntactic aspect of two sentences in calculating the similarity. Sentences are expressed as a tree of syntactic dependences. It relies on the observation that a sentence is made up of the meaning of the words that compose it as well as the syntactic links among them. The semantics of these words is evaluated on WordNet that may be problematic for the biomedical domain since WordNet does not contain specific medical terms.

Ferreira et al. [13] defined a measure to evaluate the similarity between sentences taking into account syntactic, lexical and semantic aspects of the sentence and of the words composing it. In their work the semantics of words is obtained by querying the FrameNet database and not via ontologies.

Similarity measures have also been used for ontology matching. In [28], the authors have combined three kinds of different similarity measures: lexical-based, structure-based, and semantic-based techniques as well as information in ontologies including names, labels, comments, relations and positions of concepts in the hierarchy and integrating WordNet dictionary to align ontologies.

As shown in this section, existing work rarely consider the couple LSM/ontology-based SSM to measure similarity between terms. Moreover, the only combination that we have found exploit very specific or highly generic ontologies like GO and WordNet which are not tailored to evaluate medical terms. In this work we are proposing a combination of LSM/ontology-based SSM with ontology representing the medical domain at the right level of abstraction.

4   A New Metric for Measuring Medical Term Similarity

In this section, we introduce the approach we propose to combine LSM and SSM in order to measure the similarity between medical terms. We continue with the description of the experimental setup we have defined to assess the added value of the proposed combination.

4.1   Combining Lexical and Semantic Similarity Measures

As illustrated in Sect. 3, ontology-based SSM and LSM have rarely been combined to measure the similarity between medical terms. To this end, we propose a new metric that combines ontology-based SSM and LSM as a weighted arithmetic mean, see Eq. 4. It determines the similarity between labels of two concept c_i and c_j by applying the mentioned similarity measures over two respective concepts, e.g., *C0035078:Renal failure* $\leftrightarrow$ *C0035078:Kidney failure* and attributing weights to each similarity.

In Eq. 4, the values LSM_{score} and SSM_{score} represent the normalized similarity scores given by metrics like Levenshtein and Resnik 1995 GraSM. The variables α and τ are the weights, varying in the interval of $[0.1, 1]$ with an incremental step of 0.1. It allows to change the contribution of each measure to calculate the final similarity. For instance, the configuration $\alpha = 0.8$ and $\tau = 0.3$

describe a situation where the semantic metrics are more precise than the Lexical one, but the Lexical one also contributes to the final similarity value.

$$simi(c_i, c_j) = \frac{(SSM_{score}(c_i, c_j) * \alpha) + (LSM_{score}(c_i, c_j) * \tau)}{\alpha + \tau} \qquad (4)$$

4.2   Experimental Assessment

To conduct an experimental evaluation of our new metric, we have designed a method that is based on the use of standard terminologies and existing benchmarks in order to compare our results with those generated using related work.

Terminologies

In our experiments, we have used Medical Subject Headings (MeSH) and Systematized Nomenclature of Medicine - Clinical Terms (SNOMED CT) to test the SSM. These terminologies were extracted from the UMLS. Our experiments have been done using the versions 2009AA to 2014AA (excluding the AB versions). In contrast to existing comparable approaches, we consider the evolution of concepts.

Benchmarks

We have used the three datasets suggested by [24] to evaluate our approach. We first used *MayoSRS* [31]. It contains 101 pairs of concept labels together with a score assigned to each pair denoting their relatedness. The value of the score, ranging from 0 to 10, is determined by domain experts. 0 represents a low correlation while 10 denotes a strong one.

The second dataset we have used is a subset of *MayoSRS* [31] made up of 30 pairs of concept labels. For this dataset, a distinction is made between the two categories of experts: coders and physicians and the values of the relatedness score is ranging from 1 (unrelated) to 4 (almost synonymous).

The third dataset is the UMNSRS described in [30]. Bigger than the two previous ones, it is composed of 725 concept label pairs whose similarity was evaluated by four medical experts. The similarity score of each pair was given experimentally by users based on a continuous scale ranging from 0 to 1500.

Experimental Configuration

Our aims are twofold. First, to evaluate the capacity of our approach to improve the similarity between pairs of concepts and second assess the stability of SSMs over time (with respect to the evolution of implemented ontologies). In consequence, we defined the three different configurations described hereafter:

- **Setup 1** aims at verifying the stability of semantic measures over time. To do so, we follow 3 steps: (i) we prepared the gold standard and semantic measures to be used for our experiments, e.g., dataset: *MiniMayoSRS*, SSM: *Jiang Conrath* and *i*IC: *Sanchez* (ii) we compute the similarity results using consecutive versions of MeSH and SNOMED CT and, (iii) We computed and compared Fisher's Z-Transformation to verify if the obtained results are statistically different.
- **Setup 2** verifies the number of combinations (LSM × SSM) that outperforms the single use of LSM and SSM by making α and τ vary. For this configuration we fixed the ontologies and then we grouped the results from all datasets to verify how many combinations outperformed the single measures. This setup (dataset × ontology version × measures) has produced 25920 combinations. For the sake of readability, we only highlight the overall results and the top-10 cases in the following sections.
- **Setup 3** aims at pointing out the best combinations of metrics over the three datasets. To do so, we have tested two possibilities (i) ranking with respect to the ontology. In this case, we fixed the ontology and then we analyzed the performance from all combined measures across the datasets. Here we combined all results and rank them[1]. (ii) Overall ranking regardless of the ontology and datasets. In this step we combined the previous rank and verified what measures have higher rank with lowest standard deviation.

5   Results

The results regarding the influence of ontology evolution on SSMs i.e., **setup 1**, can be observed in Table 3. For this experiments we only used the UMNSRS dataset because, among all the datasets, UMNSRS was the only one to have at least one Z-Fisher transformation value $\rho \le 0.05$, which is our threshold for considering statistical difference between the SSMs over time. The first column represent the *i*IC/SSM combination, the third column shows the versions of the ontology that have been tested. To build this column, we have considered all possible values of the set

$$\{(i,j)|i,j \in \{2009, 2010, 2011, 2012, 2013, 2014\}, i < j\}$$

The last column contains the Z-fisher transformation values obtained by comparing the computed *i*IC/SSMs and the similarity score between two terms given by domain experts.

For a sake of readability we only show in the table the combinations for which we obtained the highest Z-Fisher values (in green) as well as the lowest ones (in red). As we never obtain a value below the 0.05 threshold, we can conclude that there is no statistical difference between the value generated by any of the combination which, in turn, demonstrate a stability of Eq. 4 with respect to the used

---

[1] https://pandas.pydata.org/pandas-docs/version/0.21/generated/pandas.Series.rank.html.

Table 3. Stability of iIC/SSMs over time using UMNSRS dataset. We are considering the $\rho < 0.05$ as statistical significance. The red color indicates the lowest Z-Fisher values obtained in our experiments and the green indicates the highest ones.

iIC / SSM measures	Years	Z-fisher
Seco/Jiang Conrath	2009–2010	0.519871
Seco/Jiang Conrath	2010–2011	0.880821
Seco/Jiang Conrath	2010–2014	0.277042
Seco/Jiang Conrath	2011–2012	0.991348
Seco/Jiang Conrath	2012–2013	0.991341
Seco/Jiang Conrath	2013–2014	0.356598
Ancestors Norm/Resnik GraSM	2009–2010	0.69417
Ancestors Norm/Resnik GraSM	2010–2011	0.832429
Ancestors Norm/Resnik GraSM	2011–2012	1.0
Ancestors Norm/Resnik GraSM	2012–2013	1.0
Ancestors Norm/Resnik GraSM	2013–2014	0.793019

ontology versions. In consequence, we can conclude that SSMs are not impacted by the evolution of the underlying ontology.

Regarding **setup 2**, i.e., the percentage of combinations that outperformed the single SSMs, we observed that 5939 combinations from the 25920 possibilities (23%) outperformed the single SSMs using SNOMED CT as ontology. Concerning MeSH, only 5280 combinations from the 25920 possibilities (20%) are better. For this set of experiments, we have used the three datasets as well as all the mentioned ontology versions. This reveals a relatively low added value of the random combination of LSM and SSM with respect to the single SSM. However, when we analyzed the metrics separately, as depicted in Table 4, we can observe that for few specific combinations, the results clearly outperform the single use of SSM. This is for instance the case for the combination $AnnoMap \times Zhou/ResnikGraSM$ that is better in 91.667% of the case showing a clear added value of combining LSM and SSM. Our experiments also show that AnnoMap was the most frequent LSM that appears in the most valuable combination. The similarity computed by AnnoMap [5], see Eq. 5, is based on the combined similarity score from different string similarity functions, in particular TF/IDF, Trigram and LCS (longest common substring). The definition of AnnoMap can explain our observations.

$$sim_{AnnoMap} = MAX(TF/IDF, TriGram, LCS) \tag{5}$$

Table 5 shows combinations that do not improve the single use of SSMs at all. We observed these poor results when we combined techniques that are not complementary. For instance, Block distance, Jaccard and TF/IDF consider strings as orthogonal spaces. When combined with iIC measures focused only in the

Table 4. Percentage of combinations that outperforms the classic SSMs

LSM	iIC/SSM	%
AnnoMap	Zhou/Resnik GraSM	91.6667
	Resnik (Normalized)/Tversky iIC Ratio Model	91.6667
	Seco/Tversky iIC Ratio Model	91.6667
	Resnik (Normalized)/Resnik GraSM	87.5
	Sanchez (Normalized)/Resnik	87.5
	Seco/Resnik	87.5
	Harispe/Jiang Conrath	87.5
	Zhou/Resnik	87.5
	Seco/Resnik GraSM	87.5
	Sanchez (Normalized)/Resnik GraSM	87.5
Longest common substring	Sanchez (Normalized)/Tversky iIC Ratio Model	87.5
AnnoMap	Resnik (Normalized)/Resnik	87.5
Longest common substring	Harispe/Jiang Conrath	83.3333
LACP	Sanchez/Jian Conrath	83.3333

positioning of concepts in an ontology, the results are not improved (compared with SSMs). Note that we are not pointing good or bad techniques, but we are looking for good combination. A typical example is *Sanchez (Normalized)* that is present in both Tables 4 and 5, showing that, for instance, Block distance and Lin do not improve the outcomes, but AnnoMap and Resnik do.

Regarding **setup 3**, i.e., the overall rank for the best combinations, we experimentally verified that our approach performed better than the single SSMs regardless of the ontologies (here MeSH and SNOMED CT). We verified that the best performing combination for MeSH is (AnnoMap × Seco/Jiang Conrath) with $\alpha \in \{0.8, 1\}$ and $\tau \in \{0.4, 0.5\}$. We also observed that this combination is ranked in the top 3 best combinations but with different values for α and τ. For SNOMED CT, another combination is ranked as the most performing one. In the results the combination: (AnnoMap × Sanchez (Normalized)/Jiang Conrath) with $\alpha = 1$ and $\tau = 0.9$ was ranked first. The same behavior was observed for MeSH, where the top measure (AnnoMap × Seco/Jiang Conrath) with $\alpha = 0.8$ and $\tau = 0.5$ also appears in the top results.

The good performance of our approach is also observed when we combine all the ontologies and dataset to produce the overall rank. The final rank remains the same as we aimed at minimizing sum, average and standard deviation. In our results, we observed that (AnnoMap × Seco/Jiang Conrath) with $\alpha = 0.8$ and $\tau = 0.5$ is ranked in the top-8 in MeSH. In our experiments, the combination

Table 5. Combined measures that failed to outperform the classic ones

LSM	*i*IC/SSM
Block distance	Resnik (Normalized)/Sim *i*IC
	Sanchez (Normalized)/Lin
Levenshtein	Max Linear/Mazandu
Bigram	Ancestors Norm/Resnik
TF/IDF	Ancestors Norm/Resnik GraSM
AnnoMap	Ancestors Norm/Jiang Conrath
Jaccard	Sanchez/Jiang Conrath
	Harispe/Mazandu
Longest Common Substring	Ancestors Norm/Tversky *i*IC Ratio Norm
JaroWinkler	Ancestors Norm/Resnik GraSM
LACP	Ancestors Norm/Sim *i*IC

(AnnoMap × Seco/Jiang Conrath) with $\alpha = 0.8$ and $\tau = 0.5$ is therefore the best one.

The main difference we have observed is regarding the UMNSRS dataset, when we applied the combination (AnnoMap × Seco/Jiang Conrath) with $\alpha = 0.8$ and $\tau = 0.5$, the obtained similarity values were not greater than the single SSMs. It is due to the low Spearman's coefficient value obtained from the lexical measure $[-0.140, -0.113]$. We observed that combinations using other measures, for example, (LACP × Ancestors Norm/Lin GraSM) with $\alpha = 0.8$ and $\tau = 0.1$ show a Spearman's score of 0.462, and performs better than the single best SSM (0.456).

6   Discussion

The results of our experimental framework presented in Sect. 5 demonstrated that the combination of similarity measures, $LSM \times SSM$ formalized in Eq. 4, allows a better evaluation of medical terms relatedness. As explained in Sect. 3, very few existing work proposed to combine LSM and ontology-based SSM. In this paper, we bridge this gap by showing experimentally that the couple LSM/ontology-based SSM is of added value for measuring the similarity of medical terms. Our proposal even allow to tune the importance of both measures (LSM and SSM) with the α and τ parameters depending on the context or on the used ontologies. As a result, when calculated using single SSMs, the relatedness between *Pain* and *Morphine* (CUIs: C0030193 and C0026549) we obtain a similarity score of 0.27 but with our approach the similarity score increases to 0.56 which better correspond to the score given by domain specialists in UMNSRS dataset.

Regarding the ratio of combinations that outperformed the single SSMs, we verified that when utilizing LSMs which compare strings as a orthogonal plane, like *TF/IDF*, *Jaccard* or *Block distance*; the Spearman's Rank Correlation is low. We believe that the reason for this lies in the loss of information contained in the prefixed term, e.g., "Renal failure" ↔ "Kidney failure". When we verified the scores obtained for the MiniMayoSRS dataset, i.e., (4.0), these terms were classified as strongly related. Therefore, similarity measures should compute a higher score for this pair. However, the mentioned methods only hits a maximum similarity of 0.5 for Cosine and 0.33 for Jaccard. On the other hand, methods like LACP, provides a similarity of 0.77 that matches the scores given by the domain specialists and increases the Spearman's Rank Correlation value. Similar behavior was observed when using *Ancestors Norm* as iIC. It computes scores according to the number of ancestors from a concept divided by the total number of concepts of an ontology, i.e., $iIC = nbAncestors(v)/nbConceptInOnto$. Thus, concepts with the same number of ancestors, but in different ontology regions will have the same iIC. This limitation can be overcome if such metrics also consider sibling concepts. It plays a key role to determine the region of a concept in an ontology and is widely utilized in other domains, e.g., ontology prediction, mapping alignment as demonstrated in [10,33].

Regarding the overall rank, we observed a significant difference in the rank of the top measures for both ontologies. When we changed the dataset, the top measures substantially dropped their rank from a dataset to another. Since we verified that the three datasets do not contain many concepts having the same label, UMNSRS is the one which has the most divergence between our scores and those given by domain experts. We explain our observations as following: (i) the amount of cases to match with the domain specialties scores, around 175 in UMNSRS and 30 in the others dataset; (ii) as discussed in [30] and also verified in our experiments, the relation *similarity* → *relatedness* is directional, i.e., the terms that are similar are also related but not the opposite, e.g., the semantic similarity of *Sinemet↔Sinemet* CUIs: C0023570 and C0006982 is 0.93, while *Pain↔Morphine* CUIs: C0030193 and C0026549 is 0.27.

Finally, we verified that the used SSMs are not significantly impacted by the evolution of underlying ontologies over time. However, the size of the datasets and the number of impacted concepts they contain may moderate our conclusion. We have seen that the percentage of impacted concepts in the dataset is 2.8%, while the percentage of impacted concepts in an ontology region, i.e., *subClass*, *superClass* and *Siblings* is 5.53%. Furthermore, the top-k combinations in our overall rank, implement the measures most impacted by the ontology evolution in **setup 1**. This result highlights that the evolution of the ontologies has a role during the process of calculating the SSMs similarity. Thus, future work on semantic similarity between ontology terms has to include other pairs of impacted concepts in their dataset to verify if the stability of these measures and the obtained rank will remain the same.

7   Conclusion

In this paper, we have introduced a method that combine lexical and ontology-based semantic similarity measures to better evaluate medical terms relatedness.

We have evaluated it on three different and well-known datasets and have shown that it outperformed single use of semantic similarity measure and contribute to state-of-the-art as one of the first attempt to combine lexical and ontology-based semantic similarity measures. We also demonstrated that our proposal is not significantly affected by the evolution of underlying ontologies. In our future work, we will further evaluate our approach using larger datasets and put this metric in situation for maintaining semantic annotation impacted by ontology evolution valid over time.

References

1. Aouicha, M.B., Taieb, M.A.H.: Computing semantic similarity between biomedical concepts using new information content approach. J. Biomed. Inform. **59**, 258–275 (2016)
2. Cardoso, S.D., et al.: Leveraging the impact of ontology evolution on semantic annotations. In: Blomqvist, E., Ciancarini, P., Poggi, F., Vitali, F. (eds.) EKAW 2016. LNCS (LNAI), vol. 10024, pp. 68–82. Springer, Cham (2016). https://doi.org/10.1007/978-3-319-49004-5_5
3. Cardoso, S.D., Reynaud-Delaître, C., Da Silveira, M., Pruski, C.: Combining rules, background knowledge and change patterns to maintain semantic annotations. In: AMIA Annual Symposium, Washington DC, USA, November 2017 (2017)
4. Cardoso, S.D., et al.: Evolving semantic annotations through multiple versions of controlled medical terminologies. Health Technol. **8**, 361–376 (2018). https://doi.org/10.1007/s12553-018-0261-3
5. Christen, V., Groß, A., Varghese, J., Dugas, M., Rahm, E.: Annotating medical forms using UMLS. In: Ashish, N., Ambite, J.-L. (eds.) DILS 2015. LNCS, vol. 9162, pp. 55–69. Springer, Cham (2015). https://doi.org/10.1007/978-3-319-21843-4_5
6. Couto, F., Pinto, S.: The next generation of similarity measures that fully explore the semantics in biomedical ontologies. J. Bioinf. Comput. Biol. **11**(5), 1371001 (2013)
7. Couto, F.M., Silva, M.J., Coutinho, P.M.: Semantic similarity over the gene ontology: family correlation and selecting disjunctive ancestors. In: Proceedings of the 14th ACM International Conference on Information And Knowledge Management, pp. 343–344. ACM (2005)
8. Cross, V.: Tversky's parameterized similarity ratio model: a basis for semantic relatedness. In: 2006 Fuzzy Information Processing Society, NAFIPS 2006, Annual meeting of the North American, pp. 541–546. IEEE (2006)
9. Cross, V., Silwal, P., Chen, X.: Experiments varying semantic similarity measures and reference ontologies for ontology alignment. In: Cimiano, P., Fernández, M., Lopez, V., Schlobach, S., Völker, J. (eds.) ESWC 2013. LNCS, vol. 7955, pp. 279–281. Springer, Heidelberg (2013). https://doi.org/10.1007/978-3-642-41242-4_42
10. Da Silveira, M., Dos Reis, J.C., Pruski, C.: Management of dynamic biomedical terminologies: current status and future challenges. Yearb. Med. Inf. **10**(1), 125–133 (2015)
11. Dos Reis, J.C., Pruski, C., Da Silveira, M., Reynaud-Delaître, C.: DyKOSMap: a framework for mapping adaptation between biomedical knowledge organization systems. J. Biomed. Inf. **55**, 153–173 (2015)

12. Faria, D., Pesquita, C., Couto, F.M., Falcão, A.: Proteinon: a web tool for protein semantic similarity. Department of Informatics, University of Lisbon (2007)
13. Ferreira, R., Lins, R.D., Simske, S.J., Freitas, F., Riss, M.: Assessing sentence similarity through lexical, syntactic and semantic analysis. Comput. Speech Lang. **39**, 1–28 (2016)
14. Garla, V.N., Brandt, C.: Semantic similarity in the biomedical domain: an evaluation across knowledge sources. BMC Bioinf. **13**(1), 261 (2012)
15. Gomaa, W.H., Fahmy, A.A.: A survey of text similarity approaches. Int. J. Comput. Appl. **68**(13), 13–18 (2013)
16. Harispe, S.: Knowledge-based semantic measures: from theory to applications. Ph.D. thesis (2014)
17. Harispe, S., Sánchez, D., Ranwez, S., Janaqi, S., Montmain, J.: A framework for unifying ontology-based semantic similarity measures: a study in the biomedical domain. J. Biomed. Inf. **48**, 38–53 (2014)
18. Islam, A., Inkpen, D.: Semantic text similarity using corpus-based word similarity and string similarity. ACM Trans. Knowl. Discov. Data **2**(2), 10:1–10:25 (2008)
19. Jiang, J.J., Conrath, D.W.: Semantic similarity based on corpus statistics and lexical taxonomy. arXiv preprint https://arxiv.org/abs/cmp-lg/9709008 (1997)
20. Li, B., Wang, J.Z., Feltus, F.A., Zhou, J., Luo, F.: Effectively integrating information content and structural relationship to improve the go-based similarity measure between proteins. arXiv preprint arXiv:1001.0958 (2010)
21. Lin, D.: An information-theoretic definition of similarity. In: Proceedings of the Fifteenth International Conference on Machine Learning, ICML 1998, pp. 296–304. Morgan Kaufmann Publishers Inc., San Francisco (1998). http://dl.acm.org/citation.cfm?id=645527.657297
22. Lord, P.W., Stevens, R.D., Brass, A., Goble, C.A.: Investigating semantic similarity measures across the gene ontology: the relationship between sequence and annotation. Bioinformatics **19**(10), 1275–1283 (2003)
23. Mazandu, G.K., Mulder, N.J.: A topology-based metric for measuring term similarity in the gene ontology. Adv. Bioinform. **2012** (2012)
24. McInnes, B.T., Pedersen, T.: Evaluating measures of semantic similarity and relatedness to disambiguate terms in biomedical text. J. Biomed. Inf. **46**(6), 1116–1124 (2013). Special Section: Social Media Environments
25. Mihalcea, R., Corley, C., Strapparava, C., et al.: Corpus-based and knowledge-based measures of text semantic similarity. In: AAAI, vol. 6, 775–780 (2006)
26. Mikolov, T., Chen, K., Corrado, G., Dean, J.: Efficient estimation of word representations in vector space. arXiv preprint arXiv:1301.3781 (2013)
27. Morris, J.F.: A quantitative methodology for vetting dark network intelligence sources for social network analysis. Technical report, Air Force Inst of Tech Wright-Patterson AFB OH Graduate School of Engineering and Management (2012)
28. Nguyen, T.T., Conrad, S.: Ontology matching using multiple similarity measures. In: 2015 7th International Joint Conference on Knowledge Discovery, Knowledge Engineering and Knowledge Management (IC3K), vol. 01, pp. 603–611, November 2015. doi.ieeecomputersociety.org/
29. Oliva, J., Serrano, J.I., del Castillo, M.D., Iglesias, Á.: SyMSS: a syntax-based measure for short-text semantic similarity. Data Knowl. Eng. **70**(4), 390–405 (2011)
30. Pakhomov, S., McInnes, B., Adam, T., Liu, Y., Pedersen, T., Melton, G.B.: Semantic similarity and relatedness between clinical terms: an experimental study. In: Annual Symposium proceedings, AMIA Symposium, vol. 2010, pp. 572–576. AMIA (2010)

31. Pakhomov, S.V., Pedersen, T., McInnes, B., Melton, G.B., Ruggieri, A., Chute, C.G.: Towards a framework for developing semantic relatedness reference standards. J. Biomed. Inf. **44**(2), 251–265 (2011)

32. Pedersen, T., Pakhomov, S., Patwardhan, S., Chute, C.: Measures of semantic similarity and relatedness in the biomedical domain. J. Biomed. Inf. **40**, 288–299 (2007)

33. Pesquita, C., Couto, F.M.: Predicting the extension of biomedical ontologies. PLoS Comput. Biol. **8**(9), e1002630 (2012). https://doi.org/10.1371/journal.pcbi.1002630

34. Press, W.H., Flannery, B.P., Teukolsky, S.A., Vetterling, W.T.: Numerical Recipes in C: The Art of Scientific Computing. Cambridge University Press, New York (1988)

35. Resnik, P.: Using information content to evaluate semantic similarity in a taxonomy. In: Proceedings of the 14th International Joint Conference on Artificial Intelligence, vol. 1, pp. 448–453. Morgan Kaufmann Publishers Inc. (1995)

36. Sánchez, D., Batet, M., Isern, D.: Ontology-based information content computation. Knowl.-Based Syst. **24**(2), 297–303 (2011)

37. Sánchez, D., Batet, M., Isern, D., Valls, A.: Ontology-based semantic similarity: a new feature-based approach. Expert Syst. Appl. **39**(9), 7718–7728 (2012)

38. Seco, N., Veale, T., Hayes, J.: An intrinsic information content metric for semantic similarity in wordnet. In: ECAI, vol. 16, p. 1089 (2004)

39. Strehl, A., Ghosh, J., Mooney, R.: Impact of similarity measures on web-page clustering. In: Workshop on Artificial Intelligence for Web Search (AAAI 2000), vol. 58, p. 64 (2000)

40. Tversky, A.: Features of similarity. Psychol. Rev. **84**(4), 327 (1977)

41. Zhou, Z., Wang, Y., Gu, J.: A new model of information content for semantic similarity in wordnet. In: 2008 Second International Conference on Future Generation Communication and Networking Symposia, FGCNS 2008, vol. 3, pp. 85–89. IEEE (2008)

Construction and Visualization of Dynamic Biological Networks: Benchmarking the Neo4J Graph Database

Lena Wiese[1(✉)] [ID], Chimi Wangmo[2], Lukas Steuernagel[3], Armin O. Schmitt[3,4], and Mehmet Gültas[3,4]

[1] Institute of Computer Science, University of Göttingen, Göttingen, Germany
wiese@cs.uni-goettingen.de
[2] Gyalpozhing College of Information Technology, Royal University of Bhutan, Thimphu, Bhutan
chimiwangmo.gcit@rub.edu.bt
[3] Breeding Informatics Group, Department of Animal Sciences, University of Göttingen, Göttingen, Germany
lukas.steuernagel@stud.uni-goettingen.de,
armin.schmitt@uni-goettingen.de, gueltas@cs.uni-goettingen.de
[4] Center for Integrated Breeding Research (CiBreed), University of Göttingen, Göttingen, Germany

Abstract. Genome analysis is a major precondition for future advances in the life sciences. The complex organization of genome data and the interactions between genomic components can often be modeled and visualized in graph structures. In this paper we propose the integration of several data sets into a graph database. We study the aptness of the database system in terms of analysis and visualization of a genome regulatory network (GRN) by running a benchmark on it. Major advantages of using a database system are the modifiability of the data set, the immediate visualization of query results as well as built-in indexing and caching features.

1 Introduction

Genome analysis is a specific use case in the life sciences that has to handle large amounts of data that expose complex relationships. The size and number of genome data sets is increasing at a rapid pace [35]. Visualization of large scale data sets for exploration of various biological processes is essential to understand, e.g., the complex interplay between (bio-)chemical components or the molecular basis of relations among genes and transcription factors in regulatory networks [23]. Therefore, visualizing biological data is increasingly becoming a vital factor in the life sciences. On the one hand, it facilitates the explanation of the potential biological functions of processes in a cell-type, or the discovery of patterns as well as trends in the datasets [25]. On the other hand, visualization approaches can help researchers to generate new hypotheses to extend their knowledge based

© Springer Nature Switzerland AG 2019
S. Auer and M.-E. Vidal (Eds.): DILS 2018, LNBI 11371, pp. 33–43, 2019.
https://doi.org/10.1007/978-3-030-06016-9_3

on current informative experimental datasets and support the identification of new targets for future work [21].

Over the last decade, large efforts have been put into the visualization of biological data. For this purpose, several groups have published studies on a variety of methods and tools for e.g., statistical analysis, good layout algorithms, searching of clusters as well as data integration with well-known public repositories [1,3,8,15,18,27,28,32] (for details see review [14]). Recently, by reviewing 146 state-of-the-art visualization techniques Kerren et al. [13] have published a comprehensive interactive online visualization tool, namely BioVis Explorer, which highlights for each technique the data specific type and its characteristic analysis function within systems biology.

A fundamental research aspect of systems biology is the inference of gene regulatory networks (GRN) from experimental data to discover dynamics of disease mechanisms and to understand complex genetic programs [26]. For this aim, various tools (e.g., GENeVis [3], FastMEDUSA [4], SynTReN [5], STARNET2 [10], ARACNe [19], GeneNetWeaver [27], Cytoscape [28], NetBioV [31], LegumeGRN [32]) for the reconstruction and visualization of GRNs have been developed over the past years and those tools are widely used by system and computational biologists. A comprehensive review about (dis-)advantages of these tools can be found in [14]. Kharumnuid et al. [14] have also discussed in their review that the large majority of these tools are implemented in Java and only a few of them have been written using PHP, R, PERL, Matlab or C++, indicating that the analysis of GRNs with those tools, in most cases, needs a *two-stage process*: In the *first* stage, experimental or publicly available data from databases such as FANTOM [17], Expression Atlas [24], RNA Seq Atlas [16], or The Cancer Genome Atlas (https://www.cancer.gov/), have to be prepared; in the *second* stage, network analysis and visualization with GRN tools can be performed. This second stage possibly involves different tools for analysis and for visualization. This requires both time and detailed knowledge of tools and databases.

To overcome this limitation of existing tools as well as to simplify the construction of GRNs, we propose in this study the usage of an integrated tool, namely Neo4J, that offers both analysis as well as visualization functionality. Neo4J which is implemented in Java is a very fast, scalable graph database platform which is particularly devised for the revelation of hidden interactions within highly connected data, like complex interplay within biological systems. Further, Neo4J provides the possibility to construct *dynamic* GRNs that can be constructed and modified at runtime by insertion or deletion of nodes/edges in a stepwise progression. We demonstrate in this study that the usage of a graph database could be favourable for analysis and visualization of biological data. Especially, focusing on the construction of GRNs, it has the following advantages:

- No two-stage process consisting of a data preparation phase and a subsequent analysis and visualization phase
- Built-in disk-memory communication to load only the data relevant for processing into main memory

- Reliability of the database system with respect to long-term storage of the data (as opposed to the management of CSV files in a file system)
- Advanced indexing and caching support by the database system to speed up data processing
- Immediate visualization of analysis results even under modifications of the data set.

The article is organized as follows. Section 2 provides the necessary background on genome regulatory networks and the selection of data sets that we integrated in our study. Section 3 introduces the notion of graphs and properties of the applied graph database. Section 4 reports on the experiments with several workload queries that are applied for enhancer-promoter Interaction. Section 5 concludes this article with a discussion.

2   Data Integration

To demonstrate the usability of the Neo4J graph database for analysis and visualization of biological data in the field of life sciences, we construct GRNs based on known enhancer-promoter interactions (EPIs) and their shared regulatory processes by focusing on cooperative transcription factors (TFs). For this purpose, we first obtained biological data from different sources (FANTOM [17], UCSC genome browser [11] and PC-TraFF analysis server [21]) and then performed a mapping-based data integration process based on the following phases:

Phase 1: The information about pre-defined enhancer-promoter interactions (EPI) is obtained from the FANTOM database. FANTOM is the international research consortium for "Functional Annotation of the Mammalian Genome" that stores sets of biological data for mammalian primary cell types according to their active transcripts, transcription factors, promoters and enhancers. Using the *Human Transcribed Enhancer Atlas* in this database, we collected our benchmark data.

Phase 2: Using the UCSC genome browser, which stores a large collection of genome assemblies and annotation data, we obtained for each enhancer and promoter region (defined in Phase 1) the corresponding DNA sequences individually. It is important to note that while the sequences of enhancers are directly extracted based on their pre-defined regions, we used the annotated transcription start sites (TSS) of genes for the determination of promoter regions and extraction of their corresponding sequences (-300 base pairs to $+100$ base pairs relative to the TSS).

Phase 3: Applying the PC-TraFF analysis server to the sequences from Phase 2, we identified for each sequence a list of significant cooperative TF pairs. The PC-TraFF analysis server also provides for each TF cooperations:

- a significance score (*z-score*), which presents the strength of cooperation

– an annotation about the cooperativity of TFs—more precisely whether their physical interaction was experimentally confirmed or not. The information about their experimental validation has been obtained from TransCompel (release 2014.2) [12] and the BioGRID interaction database [6].

The data integration process for the combination of data from different sources is necessary to construct highly informative GRNs, which include complex interactions between the components of biological systems. One of the key players of these systems are the TFs which often have to form cooperative dimers in higher organisms for the effective regulation of gene expression and orchestration of distinct regulatory programs such as cell cycle, development or specificity [21,29,33]. The binding of TFs occurs in a specific combination within enhancer- and promoter regions and plays an important role in the mediation of chromatin looping, which enables enhancer-promoter interactions despite the long distances between them [2,20,22]. Today, it is well known that enhancers and promoters interact with each other in a highly selective manner through long-distance chromatin interactions to ensure coordinated cellular processes as well as cell type-specific gene expression [2,20,22]. However, it is still challenging for life scientists to understand how enhancers precisely select their target promoter(s) and which TFs facilitate such selection processes as well as interactions. To highlight such complex interactions between the elements of GRNs in a stepwise progression, Neo4J provides very effective graph database based solutions for the biological research community.

3   The Graph Database Neo4J

For datasets that lack a clear tabular structure and are of large size, data management in NoSQL databases might be more appropriate than mapping these datasets to a relational tabular format and managing them in a SQL database. Several non-relational data models and NoSQL databases—including graph data management—are surveyed in [34]. Graphs are a very versatile data model when links between entities are important. In this sense, a graph structure is also the most natural representation of a GRN.

Mathematically, a directed graph consists of a set V of nodes (or vertices) and a set E of edges. For any two nodes v_1 and v_2, a directed edge between these nodes is written as (v_1, v_2) where v_1 is the source node and v_2 is the target node. Graph databases often apply the so-called property graph data model. The property graph data model extends the notion of a directed graph by allowing key-value pairs (called "properties") to store information in the nodes and along the edges. Graph databases have been applied to several biomedical use cases in other studies: Previous versions of Neo4J have been used in a benchmark with just three queries by Have and Jensen [9] while Fiannaca et al. [7] present their BioGraphDB integration platform which is based on the OrientDB framework.

Neo4J (https://neo4j.com/) is one of the most widely used open source graph databases and has a profound community support. In Neo4J each edge has a unique type (denoting the semantics of the edge relationship between the two attached nodes); each node can have one or more labels (denoting the type or types of the node in the data model). Neo4J offers a SQL-like query language called *Cypher*. Cypher provides "declarative" syntax that is easy to read. It has an ASCII art syntax visually representing nodes and relationships in the graph structure. Thus, the query pattern for "Find all the genes g to which at least one TFPair t binds" is MATCH (g:Gene)<-[:binds]-(t:TFPair) RETURN g,t. Here, Gene and TFPair are the two types for nodes and the query identifies the relationships labeled binds connecting any nodes of type Gene and TFPair. The resulting nodes and their relationships are immediately visualized in the Neo4J browser. A snippet of the result visualization is shown in Fig. 1.

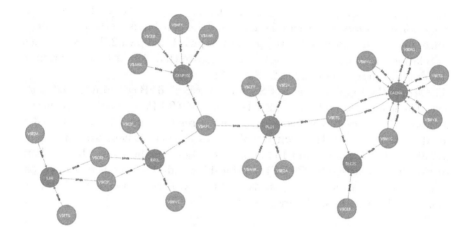

Fig. 1. A snippet of the result visualization of the sample query

Neo4j employs various caching mechanisms; as a result, once the query has been executed the following executions will use the nodes/relationships cache. The Neo4J page cache maintains data blocks in RAM for faster traversal by avoiding disk access. Moreover, the query plan cache helps reducing the computing time for parametrized queries that have already been executed before.

4   Benchmark

The benchmark was executed on a Linux PC running Ubuntu 16.04 LTS with the following specifications: an Intel CPU with 3.40 GHz and eight cores as well as 15.6 GB RAM. For our benchmark, we used Neo4J 3.4.3 Enterprise. We tested the analysis of our GRN data on a small data set and a larger data set.

4.1   Datasets

The input is provided as files in comma separated values (CSV) format. The files representing the genes (corresponding to the promoters), enhancers and pairs of transcription factors were parsed into Neo4j first. We generated three distinct types of nodes (namely, *Gene*, *Enhancer*, and *TFPair*) from them. Each *Gene* node has its *genename* as a property, while each *Enhancer* node contains a property called *enhancerID*; each *TFPair* has several properties: *pwm1*, *pwm2* (which denote the two cooperative transcription factors), the *name* (as a concatenation of the two represented transcription factors), as well as *KnownCompelPair* and *KnownBioGridPair* as properties (as described in Sect. 2, Phase 3).

Next, we created two types of relationships: *EPI* and *binds*. We extracted the *EPI* relationship between an enhancer and a promoter (located upstream of the specified gene); the EPI relationship represents the known interaction between an enhancer and promoter (as described in Sect. 2, Phase 1). The *binds* relationship links either a TFPair and an enhancer or a TFPair and a promoter. The relationship *binds* represents the fact that the pair binds to the promoter or enhancer in the order specified in the properties pwm1 and pwm2. Moreover, each *binds* relationship also has a property called *zscore* that denotes the strength of the binding (as described in Sect. 2, Phase 3).

The size of the small dataset in CSV format was 97.6 kB containing 1422 lines of text. The generated nodes included 11 genes, 619 TFPairs, and 15 enhancers; there were 19 EPI relationships and 757 binds relationships. We also tested a larger dataset of size 873 kB (with 16559 lines of text). There were 314 gene nodes, 3983 TFPair nodes, and 132 enhancer nodes. Furthermore, the numbers of relationships increased to 375 EPI relationships and 11747 binds relationships.

The datasets analyzed in this study and the cypher-commands used to load and analyze them with Neo4J are available under [30].

4.2 Queries

For both benchmark datasets, small and large, the same queries were run. The tests comprised two settings in order to consider the effects of the Neo4J cache:

- one test was conducted on cold boot and executed only once to avoid caching of the dataset;
- the other test was conducted after warming up the cache; in order to test for the real-world scenario, the queries have been run twenty times; then, their average was calculated to find the representative execution time.

The execution time represents not only the query run time on the database but includes the entire round-trip latency for visualizing the results and deserialization (streaming) of the result objects. We used the following test cases:

- Bulk data insertion
  - i1-3: Loading the CSV files (genes, enhancers, TFPairs)
  - c1-3: Assigning a uniqueness constraints to nodes
  - i4-6: Loading relationship data from CSV files (EPI and binds)
- Path queries
  - Q1a: For a given genename, find all enhancers interacting with that gene.

- Q1b: For a genename set, find all enhancers interacting with the genes.
- Q2a: For a given genename, find all TFPairs bound to that gene.
- Q2b: Restrict to the known TFPairs with AND operator.
- Q2c: Restrict to the known TFPairs with AND and OR operator
- Q2d: Find the TFPairs of an enhancer that interact with a certain gene.
- Q2e: Restrict to z-score larger than 4.
- Q3a: For all genenames find all other genenames that are bound by at least one common TFPair.
- Q3b: For a specific gene find all other genenames that are bound by at least one common TFPair.
- Q3c: For a specific enhancerID find all other enhancerIDs that are bound by at least one common TFPair.
- Q3d: For a specific enhancerID find genenames that are bound by at least one common TFPair.
- Q4a: For a given enhancer ID (or a prefix of the ID), find all the TFPairs bound to the enhancer.
- Q5a: For a given enhancerID, find all genes interacting with the enhancer.
- Q6a: For a given genename, find all TFPairs bound to the gene.
- Q6b: For a given genename, find all TFPairs bound to the gene restricting to those bindings with a high zscore.
- Q7a: For a given TF find all TFPairs that contain the TF.
- Q7b: For a given TF find the names of the two transcription factors in the TFPairs that contain the transcription factor.
- Statistical queries
  - G1a: Count the total number of TFPairs that one enhancer has in common with any other.
  - G1b: Count the TFPairs that two specific enhancers have in common.

4.3   Runtime Results

We analyzed the runtime results to assess the impact of dataset size and cache warming on our sample queries. Bulk loading data from CSV files into Neo4J is taking more time than performing any other queries as shown in Fig. 2. The increased amount of nodes in the larger benchmark (insertion steps i1, i2 and i3) did not impact the runtime substantially. In contrast, the increased amount of relationships (insertion steps i4, i5 and i6) led to a significant runtime overhead.

The next executions that cover the cold-boot tests (without cache warming) are depicted in Fig. 3. In this case, the runtime for Q2b, Q3c, Q3d, and Q5a was the same for both the small and large benchmark. Interestingly, the path queries Q1a, Q2a, Q2e Q6a, and Q6b, took on average 35% more execution time for the small benchmark than for the large benchmark which demonstrates a good off-the-shelf scalability of the graph database. Lastly, all the other queries were taking more time to execute for the large benchmarks as opposed to the small one. This overhead can be explained by the fact that the returned amount of

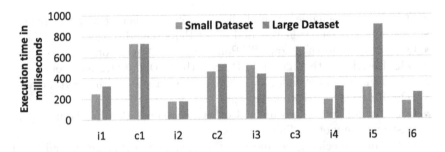

Fig. 2. Execution time for bulk data insertion steps

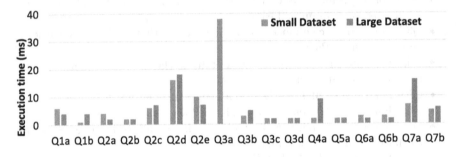

Fig. 3. Execution time of path queries on cold boot

result nodes and result relationships is significantly larger for the large benchmark. In particular, the unrestricted query Q3a (which does not provide selection conditions for the queried Genes and TFPairs) could not be executed for the large dataset because the Neo4J browser crashed after 5 min.

After warming up the cache, the performance improved drastically: the execution time for processing queries decreased by about 64% on average for both the small and the large benchmark after warming up the system as compared to execution time for the cold boot case. Notably, for both datasets the execution times are nearly similar for most of the queries, which demonstrates the positive effect of cache warming. The unrestricted query Q3a remains the exceptional case where the database is not able to finish the execution on the large data set. For some queries, in particular Q3b, Q4a, Q7a and Q7b (taking more time to execute in the large benchmark than in small benchmark) the impact of the larger result sets in the large dataset remains noticeable even after cache warming (Fig. 4).

Lastly, we tested the two COUNT queries G1a and G1b as sample queries for statistical analysis of the data sets. Here we observed a significant overhead for the larger benchmark: the first query—counting TFPairs only for each enhancer—took roughly 12 times longer for the larger benchmark (22.4 ms) than for the small benchmark (1.9 ms); more notably, the second query—counting TFPairs for each pair of enhancers—took roughly 19 times longer for the larger benchmark (37.7 ms) than for the small benchmark (1.95 ms).

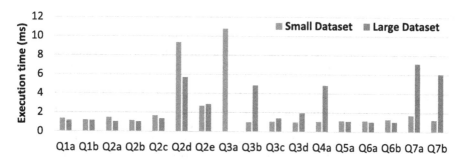

Fig. 4. Execution of path queries after cache warming

5   Conclusion

In this paper we demonstrated that several advantages can be achieved for our use case of GRN analysis by loading our data into the Neo4J graph database and expressing our analysis queries in the human-readable query language Cypher. We presented our approach for integration of biological data from different sources. We proved scalability of query execution in the graph database by benchmarking the Neo4J graph database on a query workload using a small and a large data set and investigating the effect of cache warming on the performance.

The growing importance of visualization techniques is reflected in the still growing number of corresponding publications that are registered in the Pubmed database. In 2017 the proportion of visualization related articles has increased by a factor of 17 with respect to the average from the period of 1945 to 1974. This demonstrates the drastically increasing importance of visualization techniques "in the life sciences". Up until just a few years ago publications involving the keyword visualization were typically dealing with topics related to imaging techniques in the medical sciences. Only from the year 2012 on, a substantial number of publications that deal with visualization of big data has been published.

Making big data sets accessible to interpretation is one of the main challenges in Life science now and in the next years. Graph databases (in particular Neo4J) can be a powerful tool to aid researchers with the storage, the integration as well the analysis and visualization of biological, medical and healthcare data.

Acknowledgements. Chimi Wangmo participated in the preparation of this article while visiting the University of Göttingen with a Go International Plus scholarship by the Erasmus+ Key Action of the European Commission.

References

1. Albers, D., Dewey, C., Gleicher, M.: Sequence surveyor: leveraging overview for scalable genomic alignment visualization. IEEE Trans. Vis. Comput. Graph. **17**(12), 2392–2401 (2011)
2. van Arensbergen, J., van Steensel, B., Bussemaker, H.J.: In search of the determinants of enhancer-promoter interaction specificity. Trends Cell Biol. **24**(11), 695–702 (2014). http://www.sciencedirect.com/science/article/pii/S0962892414001184
3. Baker, C.A., Carpendale, M.S.T., Prusinkiewicz, P., Surette, M.G.: GeneVis: visualization tools for genetic regulatory network dynamics. In: Visualization, VIS 2002. IEEE. pp. 243–250 (2002)
4. Bozdag, S., Li, A., Wuchty, S., Fine, H.A.: FastMEDUSA: a parallelized tool to infer gene regulatory networks. Bioinformatics **26**(14), 1792–1793 (2010)
5. Van den Bulcke, T., et al.: SynTReN: a generator of synthetic gene expression data for design and analysis of structure learning algorithms. BMC Bioinform. **7**(1), 43 (2006). https://doi.org/10.1186/1471-2105-7-43
6. Chatraryamontri, A., et al.: The BioGRID interaction database: 2015 update. Nucleic Acids Res. (2014). http://nar.oxfordjournals.org/content/early/2014/11/26/nar.gku1204.abstract
7. Fiannaca, A., La Rosa, M., La Paglia, L., Messina, A., Urso, A.: BioGraphDB: a new graphDB collecting heterogeneous data for bioinformatics analysis. In: Proceedings of BIOTECHNO (2016)
8. Gomez, J., et al.: BioJS: an open source Javascript framework for biological data visualization. Bioinformatics **29**(8), 1103–1104 (2013). https://doi.org/10.1093/bioinformatics/btt100
9. Have, C.T., Jensen, L.J.: Are graph databases ready for bioinformatics? Bioinformatics **29**(24), 3107 (2013)
10. Jupiter, D., Chen, H., VanBuren, V.: STARNET2: a web-based tool for accelerating discovery of gene regulatory networks using microarray co-expression data. BMC Bioinform. **10**(1), 332 (2009)
11. Karolchik, D., et al.: The UCSC table browser data retrieval tool. Nucleic Acids Res. **32**(suppl-1), D493–D496 (2004). https://doi.org/10.1093/nar/gkh103
12. Kel-Margoulis, O., Kel, A., Reuter, I., Deineko, I., Wingender, E.: TRANSCompel: a database on composite regulatory elements in eukaryotic genes. Nucleic Acids Res. **30**, 332–334 (2002)
13. Kerren, A., Kucher, K., Li, Y.F., Schreiber, F.: Biovis explorer: a visual guide for biological data visualization techniques. PLOS ONE **12**(11), 1–14 (2017). https://doi.org/10.1371/journal.pone.0187341
14. Kharumnuid, G., Roy, S.: Tools for in-silico reconstruction and visualization of gene regulatory networks (GRN). In: 2015 Second International Conference on Advances in Computing and Communication Engineering (ICACCE), pp. 421–426. IEEE (2015)
15. Kirlew, P.W.: Life science data repositories in the publications of scientists and librarians. Issues Sci. Technol. Libr. **65** (2011)
16. Krupp, M., Marquardt, J.U., Sahin, U., Galle, P.R., Castle, J., Teufel, A.: RNA-Seq Atlas - a reference database for gene expression profiling in normal tissue by next-generation sequencing. Bioinformatics **28**(8), 1184–1185 (2012). https://doi.org/10.1093/bioinformatics/bts084
17. Lizio, M., et al.: Gateways to the FANTOM5 promoter level mammalian expression atlas. Genome Biol. **16**(1), 22 (2015). https://doi.org/10.1186/s13059-014-0560-6

18. Longabaugh, W.J., Davidson, E.H., Bolouri, H.: Visualization, documentation, analysis, and communication of large-scale gene regulatory networks. Biochim. Biophys. Acta (BBA) - Gene Regul. Mech. **1789**(4), 363–374 (2009). http://www.sciencedirect.com/science/article/pii/S1874939908001624

19. Margolin, A.A., et al.: ARACNE: an algorithm for the reconstruction of gene regulatory networks in a mammalian cellular context. BMC Bioinform. **7**(1), S7 (2006)

20. Matharu, N., Ahituv, N.: Minor loops in major folds: enhancer-promoter looping, chromatin restructuring, and their association with transcriptional regulation and disease. PLOS Genet. **11**(12), 1–14 (2015). https://doi.org/10.1371/journal.pgen.1005640

21. Meckbach, C., Tacke, R., Hua, X., Waack, S., Wingender, E., Gültas, M.: PC-TraFF: identification of potentially collaborating transcription factors using pointwise mutual information. BMC Bioinform. **16**(1), 400 (2015). https://doi.org/10.1186/s12859-015-0827-2

22. Mora, A., Sandve, G.K., Gabrielsen, O.S., Eskeland, R.: In the loop: promoter-enhancer interactions and bioinformatics. Brief. Bioinform. **17**(6), 980–995 (2016). https://doi.org/10.1093/bib/bbv097

23. O'Donoghue, S.I., et al.: Visualizing biological data - now and in the future. Nature Methods **7**(3), S2 (2010)

24. Petryszak, R., et al.: Expression Atlas update - an integrated database of gene and protein expression in humans, animals and plants. Nucleic Acids Res. **44**(D1), D746–D752 (2015)

25. Ren, J., Lu, J., Wang, L., Chen, D.: Data visualization in bioinformatics. Adv. Inf. Sci. Serv. Sci. **4**(22) (2012)

26. Roy, S., Bhattacharyya, D.K., Kalita, J.K.: Reconstruction of gene co-expression network from microarray data using local expression patterns. BMC Bioinform. **15**(7), S10 (2014)

27. Schaffter, T., Marbach, D., Floreano, D.: Genenetweaver: in silico benchmark generation and performance profiling of network inference methods. Bioinformatics **27**(16), 2263–2270 (2011)

28. Smoot, M.E., Ono, K., Ruscheinski, J., Wang, P.L., Ideker, T.: Cytoscape 2.8: new features for data integration and network visualization. Bioinformatics **27**(3), 431–432 (2010)

29. Sonawane, A.R., et al.: Understanding tissue-specific gene regulation. Cell Rep. **21**(4), 1077–1088 (2017)

30. Steuernagel, L., Wiese, L., Gültas, M.: Repository visualization of dynamic biological networks. https://github.com/azifiDils/Visualization-of-DynamicBiological-Networks-

31. Tripathi, S., Dehmer, M., Emmert-Streib, F.: NetBioV: an R package for visualizing large network data in biology and medicine. Bioinformatics **30**(19), 2834–2836 (2014)

32. Wang, M., et al.: LegumeGRN: a gene regulatory network prediction server for functional and comparative studies. PloS One **8**(7), e67434 (2013)

33. Whitfield, T.W., et al.: Functional analysis of transcription factor binding sites in human promoters. Genome Biol. **13**(9), R50 (2012). https://doi.org/10.1186/gb-2012-13-9-r50

34. Wiese, L.: Advanced Data Management for SQL, NoSQL. Cloud and Distributed Databases, DeGruyter/Oldenbourg (2015)

35. Wiese, L., Schmitt, A.O., Gültas, M.: Big data technologies for DNA sequencing. In: Sakr, S., Zomaya, A. (eds.) Encyclopedia of Big Data Technologies. Springer, Cham (2018). https://doi.org/10.1007/978-3-319-63962-8

A Knowledge-Driven Pipeline
for Transforming Big Data
into Actionable Knowledge

Maria-Esther Vidal[1,2] (iD), Kemele M. Endris[1,2] (iD), Samaneh Jozashoori[1,2](✉) (iD),
and Guillermo Palma[1,2] (iD)

[1] TIB Leibniz Information Centre for Science and Technology, Hannover, Germany
maria.vidal@tib.eu, {endris,jozashoori,palma}@l3s.de
[2] L3S Institute, Leibniz University of Hannover, Hannover, Germany

Abstract. Big biomedical data has grown exponentially during the last
decades, as well as the applications that demand the understanding and
discovery of the knowledge encoded in available big data. In order to
address these requirements while scaling up to the dominant dimensions
of big biomedical data –volume, variety, and veracity– novel data integration techniques need to be defined. In this paper, we devise a knowledge-
driven approach that relies on Semantic Web technologies such as ontologies, mapping languages, linked data, to generate a knowledge graph
that integrates big data. Furthermore, query processing and knowledge
discovery methods are implemented on top of the knowledge graph for
enabling exploration and pattern uncovering. We report on the results
of applying the proposed knowledge-driven approach in the EU funded
project iASiS (http://project-iasis.eu/). in order to transform big data
into actionable knowledge, paying thus the way for precision medicine
and health policy making.

1 Introduction

Big data plays an important role in promoting sustained economic growth of
countries and companies through industrial digitization, and emerging scientific and interdisciplinary research. Specifically, significant contributions have
been achieved by conducting big data-driven studies over clinical and genomic
data with the aim of supporting precision medicine [11]. Exemplary contributions include big data analytics over Electronic Health Records (EHRs) of nearly
three million people and trillions of pieces of medical data for identifying associations between the use of proton-pump inhibitors and the likelihood of incurring
a heart attack [12]. Despite the significant impact of big data, we are entering
into a new era where domains like genomic, are projected to grow very rapidly
in the next decade, reaching more than one Zetta bytes of heterogeneous data
per year by 2025 [14]. In this next era, transforming big data into actionable
big knowledge will require novel and scalable tools for enabling not only big

S. Auer and M.-E. Vidal (Eds.): DILS 2018, LNBI 11371, pp. 44–49, 2019.
https://doi.org/10.1007/978-3-030-06016-9_4

data ingestion and curation, but also for efficient large-scale knowledge extraction, integration, exploration, and discovery. In this poster paper, we describe a knowledge-driven pipeline devised with the aim of addressing these challenges. The pipeline resorts to text mining, image processing methods, and ontologies to extract knowledge encoded in unstructured Big data and to describe extracted knowledge with terms from ontologies. Then, extracted knowledge is integrated into a knowledge graph. A unified schema is used to describe and structure the extracted in the knowledge graph. Annotations from ontologies provide the basis for data integration and for linking integrated data with equivalent concepts in existing knowledge graphs. Finally, knowledge discovery is performed by exploring and analyzing the knowledge graph. The proposed knowledge-driven approach is being utilized to integrate biomedical data, e.g., drugs, genes, mutations, side effects, with clinical records, medical images, and geneomic data. As a result, a knowledge graph with more than 250 million RDF triples has been created. Albeit initial, this knowledge graph enables the discovery of patterns that could not be found in raw data. Patterns include mutations that impact on the effectiveness of a drug, side-effects of a drug, and drug-target interactions.

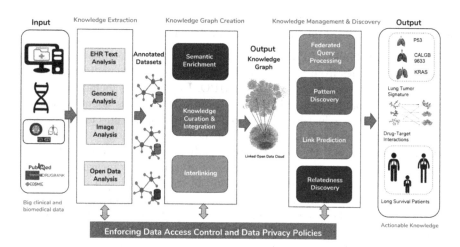

Fig. 1. A Knowledge-Driven Pipeline. Heterogeneous data sources are received as input, and a knowledge graph and unknown patterns and relations are output. The knowledge graph is linked to related knowledge graphs; federated query processing and knowledge discovery techniques enable knowledge exploration and discovery. Data privacy and access regulations imposed by data providers are enforced.

2   A Knowledge-Driven Pipeline

Our knowledge-driven pipeline receives big data sources in different formats, e.g., clinical notes, images, scientific publications, and structured data. It generates a knowledge graph from which unknown patterns and relationships can be discovered; Fig. 1 depicts the following main components of the pipeline:

EHR Text Analysis: Semi-automatic data curation techniques are utilized for data quality assurance, e.g., removing duplicates, solving ambiguities, and completing missing attributes. Natural Language Processing (NLP) techniques are applied to extract relevant entities from unstructured fields, i.e., clinical notes or lab test results. NLP techniques rely on medical vocabularies, e.g., Unified Medical Language System (UMLS)[1] or Human Phenotype Ontology (HPO)[2], NLP corpuses and tools, e.g., lemmatization or Named Entity Recognition, to annotate concepts with terms from medical vocabularies.

Genomic Analysis: Data mining tools, e.g., catRapid [7], are applied to identify protein-RNA associations with high accuracy. Publicly available datasets, e.g., data from GTEx, GEO, and ArrayExpress, are used for the integration with transcriptomic data. Finally, this component relies on the Gene Ontology to determine key genes for lung cancer and interactions between these genes. Furthermore, genes are annotated with identifiers from different databases, e.g., HUGO or Uniprot/SwissProt, as well as Human Phenotype Ontology (HPO).

Image Analysis: Machine learning algorithms are employed to learn predictive models able to classified medical images and detect lung tumors.

Open Data Analysis: NLP and network analysis methods enable the semantic annotation of entities from biomedical data sources using biomedical ontologies and medical vocabularies, e.g., UMLS or HPO. Data sources include PubMed[3], COSMIC[4], DrugBank[5], and STITCH[6]. Annotated datasets comprise entities like mutations, genes, scientific publications, biomarkers, side effects, transcripts, proteins, and drugs, as well as relations between these entities.

A knowledge graph is created by semantically describing entities using a unified schema. Annotations are exploited by semantic similarity measures [10] with the aim of determining relatedness between the entities included in the knowledge graph, as well as for duplicate and inconsistency detection. Related entities are integrated into the knowledge graph following different fusion policies [3]. Fusion policies resemble flexible filters tailored for specific tasks, e.g., keep all literals with different language tags or retain an authoritative value; replace one attribute with another; merge all the attributes of an entity in the knowledge graph; etc. Ontological axioms of the dataset annotations are fired for resolving conflicts and inequalities during the evaluation of the fusion policies. Entities in the knowledge graph are linked to equivalent entities in knowledge graphs in the Linked Open Data Cloud. Linking techniques resort to semantic similarity metrics and the semantic encoded in the ontologies of the different knowledge

---

[1] https://www.nlm.nih.gov/research/umls/.

[2] https://hpo.jax.org/app/.

[3] https://www.ncbi.nlm.nih.gov/pubmed/.

[4] https://cancer.sanger.ac.uk/cosmic.

[5] https://www.drugbank.ca/.

[6] http://stitch.embl.de/.

graphs, for determining when entities in different knowledge graphs, e.g., mutations and genes in TCGA-A[7]. Knowledge represented in the knowledge graph
and links to other knowledge graphs, is explored by a federated query processing
engine, and knowledge discovery methods are used to uncover patterns in the
knowledge graphs. Finally, data privacy and access controlled regulations are
enforced during the execution of the tasks of the pipeline [4].

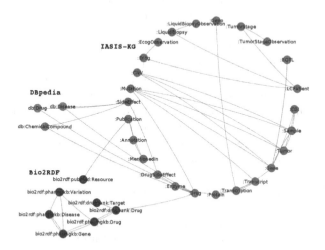

Fig. 2. Connectivity of IASIS-KG. Graph representing the connectivity of the RDF
classes in IASIS-KG, and DBpedia and Bio2RDF. All the RDF classes are connected.

3   Initial Results

Following the proposed knowledge-driven pipeline, data from twelve datasets has
been integrated. A unified schema allows for data description in a knowledge
graph; it includes 49 classes, 56 `ObjectProperty`, and 74 `DatatypeProperty`.
The number of properties per class in the unified schema ranges from five to
80; the majority of the classes have less than 10 properties, and classes with
a higher number of properties correspond to superclasses which inherit all the
properties of their subclasses. The process of graph creation enables the creation
of a knowledge graph with 236,512,819 RDF triples, 26 RDF classes, and in
average, 6.98 properties per entity; it is named as IASIS-KG. In average there
are 86,934.00 entities per RDF class, some RDF classes may have up to 20
million entities. Figure 2 shows the connectivity between the RDF classes in
IASIS-KG; there are 35 nodes in the graph, while 58 edges represent links among
RDF classes. Also, it can be observed that all the RDF classes are connected
to at least one RDF class, i.e., there are no isolated classes. These statistics
facilitate the understanding of the amount of represented knowledge, as well as
the opportunities offered by IASIS-KG for knowledge exploration and discovery.

---

[7] http://tcga.deri.ie/.

4   Related Work

Biomedical datasets are characterized by the "Vs" challenges of big data, i.e., volume, velocity, variety, veracity, value, and variability [13]. To address the data complexity issues imposed by these challenges, novel paradigms and technologies have been proposed in the last years. Exemplary platforms include the BigDataEurope platform [1], an easy-to-deploy architecture that combines technologies to process large and heterogeneous sources. An extensive literature analysis on big data methods [13]indicates that the state of the art focuses on specific dimensions of data complexity, whereas isolated solutions are not sufficient to meet the demands imposed by the transformation of big data into actionable knowledge [15]. In order to represent the meaning of biomedical entities several ontologies and controlled vocabularies have been defined, e.g., HPO and UMLS. These ontologies are commonly utilized to provide a unique representation of concepts extracted from unstructured or structured datasets [9]. Likewise, knowledge graphs are especially important in knowledge representation, because they provide a common knowledge structure to integrate and semantically describe the meaning of entities from diverse domains. Generic knowledge graphs like DBpedia [6] and Yago [8], or) describe generic facts, e.g., persons, organizations, or cities, while more specific knowledge graphs like KnowLife [5] and Bio2RDF [2] exploit domain specific vocabularies like UMLS to integrate biomedical data items like publications, genes, mutations, drugs, and diseases. Similarly, the proposed knowledge-driven approach relies on semantic annotations from ontologies, e.g., HPO and UMLS. However, in contrast to existing approaches, these annotations are used as building blocks for the semantic integration process and well as curation. Thus, this solution is able to scale up to the veracity and variety characteristics of the collected heterogeneous biomedical.

5   Conclusions

A knowledge-driven pipeline for transforming Big data into a knowledge graph is presented; it comprises components that enable knowledge extraction, a knowledge graph creation, and knowledge management and discovery. As a proof of concept, the proposed pipeline has been applied in the context of the European Union Horizon 2020 funded project iASiS. As a result, a knowledge graph with more than 230 million RDF triples have been created. This knowledge graph includes mutations that impact on the effectiveness of a drug, side-effects of a drug, and drug-target interactions, and represents a building block for the exploration and discovery of potential novel patterns. Furthermore, initial results illustrate the feasibility of the approach, as well as the relevant role of Semantic Web technologies and ontologies in the process of data integration. In the future, this pipeline will be used in other biomedical use cases, and novel machine learning approaches over the knowledge graph will be implemented.

Acknowledgement. This work has been supported by the European Union's Horizon 2020 Research and Innovation Program for the project iASiS with grant agreement No 727658.

References

1. Auer, S., et al.: The bigdataeurope platform - supporting the variety dimension of big data. In: Web Engineering - 17th International Conference, ICWE 2017, pp. 41–59 (2017)
2. Belleau, F., Nolin, M., Tourigny, N., Rigault, P., Morissette, J.: Bio2RDF: towards a mashup to build bioinformatics knowledge systems. J. Biomed. Inform. **41**(5), 706–716 (2008)
3. Collarana, D., Galkin, M., Ribón, I.T., Vidal, M., Lange, C., Auer, S.: MINTE: semantically integrating RDF graphs. In Proceedings of the 7th International Conference on Web Intelligence, Mining and Semantics, WIMS 2017, Amantea, Italy, 19–22 June 2017 (2017)
4. Endris, K.M., Almhithawi, Z., Lytra, I., Vidal, M., Auer, S.: BOUNCER: privacy-aware query processing over federations of RDF datasets. In: Database and Expert Systems Applications - 29th International Conference, DEXA 2018, Regensburg, Germany, 3–6 September 2018, Proceedings, Part I, pp. 69–84 (2018)
5. Ernst, P., Siu, A., Weikum, G.: Knowlife: a versatile approach for constructing a large knowledge graph for biomedical sciences. BMC Bioinform. **16**, 157 (2015)
6. Lehmann, J., et al.: DBpedia - a large-scale, multilingual knowledge base extracted from Wikipedia. Semant. Web **6**(2), 167–195 (2015)
7. Livi, C.M., Klus, P., Delli Ponti, R., Tartaglia, G.G: cat Rapid signature: identification of ribonucleoproteins and RNA-binding regions. Bioinformatics **32**(5), 773–775 (2016)
8. Mahdisoltani, F., Biega, J., Suchanek, F.M.: YAGO3: a knowledge base from multilingual Wikipedias. In CIDR 2015 (2015)
9. Menasalvas, E., Rodriguez-Gonzalez, A., Costumero, R., Ambit, H., Gonzalo, C.: Clinical narrative analytics challenges. In: Flores, V., et al. (eds.) IJCRS 2016. LNCS (LNAI), vol. 9920, pp. 23–32. Springer, Cham (2016). https://doi.org/10. 1007/978-3-319-47160-0_2
10. Ribón, I.T., Vidal, M., Kämpgen, B., Sure-Vetter, Y.: GADES: a graph-based semantic similarity measure. In: Proceedings of SEMANTICS, pp. 101–104 (2016)
11. Schmidlen, T.J., Wawak, L., Kasper, R., García-España, J.F., Christman, M.F., Gordon, E.S.: Personalized genomic results: analysis of informational needs. J. Genetic Counseling **23**(4), 578–587 (2014)
12. Shah, N.H., et al.: Proton pump inhibitor usage and the risk of myocardial infarction in the general population. Plos One **10**(7), e0124653 (2015)
13. Sivarajah, U.M.M.K., Irani, Z., Weerakkody, V.: Critical analysis of big data challenges and analytical methods. J. Bus. Res. **70**, 263–286 (2017)
14. Stephens, Z.D., et al.: Big data: astronomical or genomical? Plos One **13**(7), e1002195 (2015)
15. Jagadish, H.V., et al.: Big data and its technical challenges. Commun. ACM **57**(7), 86–94 (2014)

Leaving No Stone Unturned: Using Machine Learning Based Approaches for Information Extraction from Full Texts of a Research Data Warehouse

Johanna Fiebeck[1]([⊠]) (iD), Hans Laser[1] (iD), Hinrich B. Winther[2] (iD),
and Svetlana Gerbel[1]([⊠]) (iD)

[1] Centre for Information Management, Hannover Medical School,
Carl-Neuberg-Str. 1, 30625 Hannover, Germany
{fiebeck.johanna, gerbel.svetlana}@mh-hannover.de
[2] Institute for Diagnostic and Interventional Radiology,
Hannover Medical School, Carl-Neuberg-Str. 1, 30625 Hannover, Germany

Abstract. Data in healthcare and routine medical treatment is growing fast. Therefore and because of its variety, possible correlation within these are becoming even more complex. Popular tools for facilitating the daily routine for the clinical researchers are more often based on machine learning (ML) algorithms. Those tools might facilitate data management, data integration or even content classification. Besides commercial functionalities, there are many solutions which are developed by the user himself for his own, specific question of research or task. One of these tasks is described within this work: qualifying the Weber fracture, an ankle joint fracture, from radiological findings with the help of supervised machine learning algorithms. To do so, the findings were firstly processed with common natural language processing (NLP) methods. For the classifying part, we used the bags-of-words-approach to bring together the medical findings on the one hand, and the metadata of the findings on the other hand, and compared several common classifier to have the best results. In order to conduct this study, we used the data and the technology of the Enterprise Clinical Research Data Warehouse (ECRDW) from Hannover Medical School. This paper shows the implementation of machine learning and NLP techniques into the data warehouse integration process in order to provide consolidated, processed and qualified data to be queried for teaching and research purposes.

Keywords: Clinical Research Data Warehouse · Machine learning
Text mining · Data science · Unstructured data · Secondary use
Radiology · NLP

1 Introduction

Medical records, pathology and radiology findings or medication are often available in an unstructured form. Relevant information is therefore not always described in concrete fields, but mostly in free text form. Drawing inferences on disease progressions, processes or statistics for quality assurance are difficult to extract from this information.

© Springer Nature Switzerland AG 2019
S. Auer and M.-E. Vidal (Eds.): DILS 2018, LNBI 11371, pp. 50–58, 2019.
https://doi.org/10.1007/978-3-030-06016-9_5

The structure of the texts differs within departments and partly between findings. Machine Learning (ML) methods can be used to solve this problem. Frequent data mining tasks in radiology include [1]:

- Automated derivation of numbers for defined instances or from finding results (feature extraction) from the unstructured text [2]
- Information enrichment of structured data by feature extraction
- Text analysis using controlled terminologies, in radiology (mainly the terminology RadLex [1, 3])
- Classification and clustering, e.g. to identify patient cohorts (selection of patients with similar clinical pictures) [4]

At the Hannover Medical School (MHH) the radiological findings are captured in the Radiology Information System (RIS) in free text form but are divided into individual, predefined sections.

In this study, we used these semi-structured findings data integrated into the MHH Enterprise Clinical Research Data Warehouse (ECRDW). The ECRDW of the MHH is an interdisciplinary data integration and analysis platform for research-relevant issues. In the clinical-university sector, a data warehouse based technology, serves to consolidate data routinely generated in health care for secondary use purposes [5, 6]. The typical use cases of a clinical data warehouse (CDW) include:

- Patient screening for clinical trials
- Epidemiological estimations
- Validation of data in research databases and their data enrichment with the aim of quality improvement in research tasks
- Development of decision support approaches for specific research questions

In order to make these findings available for queries, a method for data cleansing and data processing is to be developed and implemented within the standard ETL (Extraction, Transformation, and Loading) process of the ECRDW.

Our research question is to locate radiology findings that refer to the so-called Weber fracture, an ankle fracture, in order to be able to analyze the corresponding X-ray images or to make them available for teaching courses. To do so, the findings are to be preprocessed in a structured manner with the aid of natural language processing (NLP) methods and to classify the records with ML algorithms. We decided to use ML methods because the diagnosis often is not exactly named in the text. Thus, a simple full text search will not find all relevant results or will also find negating results. Additionally search for possible synonyms is required. By using ML techniques, we also included some report metadata as features, such as radiology service group and department.

This paper is divided into the typical sections Materials and methods, results, and discussion and conclusion. In Materials and methods the necessary steps for the preprocessing of the relevant data and the structure of the ML pipeline are described. In the section results the particularities of the original data set are first summarized and then the resulting selection of a suitable algorithm for ML with corresponding metrics explained. Subsequently, the result of the prediction of the Weber fracture and the implementation of the process into the ETL process are outlined.

2 Materials and Methods

2.1 Accessing the Data via a Data Warehouse Plattform

In order to develop appropriate methods for information retrieval technology and data of the MHH ECRDW was used. The ECRDW is based on the Microsoft (MS) SQL Server Stack. The basis for the machine learning are radiological findings (with additional metadata), which were joined with ICD10-GM[1] diagnosis codes. The metadata and diagnosis codes were used to select necessary features and annotate the training data on the one hand and to access the medical records for prediction on the other. A total of 2,000 medical findings were identified for training and 17,354 medical findings were predicted on the basis of the following diagnosis codes:

- S82 as fracture of the lower leg, including the upper ankle joint
- S82.5 as fracture of the inner ankle
- S82.6 as fracture of the outer ankle
- S92 as fracture of the foot (except upper ankle)

The findings from the RIS were integrated into the ECRDW via HL7 by using the MS integration services. By doing so, the whole semi-structured finding text is integrated as a full text, while the findings are split into four separated fields: "Klinische Angaben" (engl. "clinical data"), "Fragestellung" (engl. "clinical situation"), "Befund" (engl. "finding"), "Beurteilung" (engl. "assessment"). Within this text the headings are displayed by pseudo-html tags like /.br.//. For re-separating the fields, we developed a regular expression (regex) term addressing these tags. By doing so, this regex also may search for further headings which might be feasible for integrating in clinical routine. Thus, the final regex is:

$$(\text{"}\backslash S\backslash.\text{br}\backslash S\backslash w+[:]\backslash S[.]\backslash w+\backslash S\text{"})$$

The regex and the NLP pipeline was implemented by using the natural language toolkit Python NLTK for general text processing [7, 8].

The NLP pipeline for splitting the text into their pre-defined sections includes several steps such as:

- Loading the data
- Hard-coded misspelling-cleansing
- Extracting the headings with the regex and export the top 10-headings
- Splitting the text according to their predefined headings into predefined fields.

2.2 Machine Learning Pipeline for Classification

The ML training data consists of 2,000 randomly selected radiology findings. The potential radiology findings have already been selected by ICD codes (range of lower leg injuries). To create a binary classification, the full text of the ICD diagnosis was

---

[1] ICD-GM: "International Classification of Diseases, German Modification" is the official classification for diagnoses in outpatient and inpatient health care in Germany.

searched for the keyword "Weber". In addition, the service type "ankle joint", which was documented in the metadata of the radiological findings, served as a further criterion to create the positive class. The negative class was created by using radiological findings not having these specific attributes.

The findings then were prepared in a machine learning pipeline, consisting of a text preprocessing part and a classification part. The text preprocessing steps for both the training data and the data to be classified were carried out with the Python NLTK packages as well. These included tokenizing, removing stopwords, transformation into bags of words and converting the bags of words into a Python Pandas DataFrame.

For dimension reduction within the bags of words, we inserted the following steps and compared the results:

- Filtering dates and times using the RegexpTokenizer (optional step)
- Only the twenty-most often tokens where represented in the bags of words, using the built-in Counter function.

Afterwards, we joined the bags-of-words dataframe with metadata features: service group, service type, analysis device, operating department. The prepared training data was used for classifier training. Various ML classifier were selected for comparision: Naive Bayes Classifier, Support Vector Machines, Decision Trees and Random Forest and Logistic Regression algorithms and accessed them via the Python scikit-learn package. We chose a 10-fold cross-validation with 70/30 training/test split.

Additionally, selected algorithms were judged according their confusion matrices while predicting test data. "Unknown" medical records for prediction where read directly from the ECRDW once they are in the same ICD range as selected above.

Prediction was conducted on the medical record as well as on the section "Befund" ("medical indication") only. In Table 1, the process of prediction is outlined.

Table 1. Steps in machine learning pipeline from loading and preparing the data to prediction and analysis.

Step	Description
1	Selecting and annotating training data
2	Feature selection
3	Loading unknown data for prediction
4	NLP pipeline
5	Reduction of *bags-of-words* dimensions of the "unknown" data down to the dimensions of the training data set
6	Training of the algorithm, 10-fold cross-validation (70/30 split)
7	Prediction on test data with confusion matrix
8	Classification of unknown medical records
9	Embedding the prediction results to the data

3 Results

3.1 Finding Patterns and Separating the Text

As expected, the headings of the predefined fields were found most often, but not all of them actually were filled (Fig. 1). Additionally to the predefined headings, the regex found some potential new headings, for example "Methodik" ("methods"), which is named in about 5% of all findings. The other headings found by the regex are mainly indicating anatomical issues. By these results, one can recommend to add the "methods" as an additional section within the radiological findings.

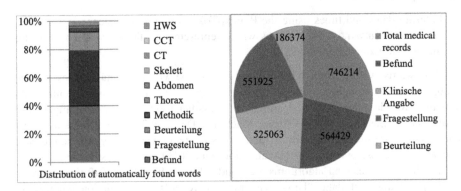

Fig. 1. Left side: the 10 most-often headings extracted from medical record texts, plotted according their occurrences. Right side: filled section quantity by their pre-defined headings.

3.2 Finding the Appropriate Machine Learning Algorithm

After comparing the accuracies in 10-fold cross-validation of all algorithms and hyperparameter modes (Table 2), we selected two algorithms for classification of unknown findings: the standard Decision Tree and the standard Support Vector Machine (SVM) algorithm.

Additionally, we decided to add the RegExpTokenizer for dimension reduction. Although there was only minor effect in the accuracies whether the additional tokenizer was conducted or not, it helped to reduce feature dimensions drastically.

After selecting the algorithms, they were chosen to predict test data within a confusion matrix. As shown in Fig. 2, both prediction results vary seriously from each other. As expected, the Decision Tree algorithm performs slightly better than the SVM algorithm: While the correct-positive rate and the correct-negative rate of the Decision Tree is quite high (171:174), the SVM predicts a high rate of correct-positive records but a very high false-positive class (197:140).

Table 2. Results of comparing several classification algorithms accuracies in 10-fold cross-validation, 70/30 split mode. As there 1,000 positive and 1,000 negative training data, the baseline indicator is set to 50%.

Modus	Without RegexpTokenizer		With RegexpTokenizer	
Standard SVM	54.00%	(+/− 0.08)	54.00%	(+/− 0.08)
SVM (kernel = "linear", C = 0.025)	54.00%	(+/− 0.08)	54.00%	(+/− 0.08)
SVM (gamma = 2, C = 1)	58.00%	(+/− 0.04)	58.00%	(+/− 0.04)
Standard Decision Tree	70.00%	(+/− 0.08)	70.00%	(+/− 0.08)
Decision Tree (max_depth = 5)	58.00%	(+/− 0.08)	59.00%	(+/− 0.07)
Standard Random Forest	70.00%	(+/− 0.09)	70.00%	(+/− 0.09)
Random Forest (max_depth = 5, n_estimators = 10, max_features = 2)	50.00%	(+/− 0.04)	51.00%	(+/− 0.02)
Naive Bayes_Gaussian	54.00%	(+/− 0.07)	54.00%	(+/− 0.07)
Naive Bayes_BernoulliNB	50.00%	(+/− 0.03)	50.00%	(+/− 0.03)
Naive Bayes_MultinomialNB	55.00%	(+/− 0.08)	55.00%	(+/− 0.08)
Standard Logistic Regression	54.00%	(+/− 0.09)	54.00%	(+/− 0.09)

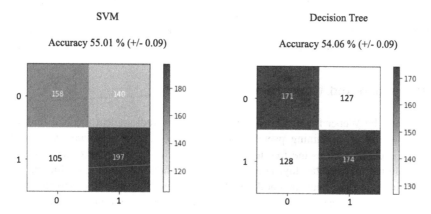

Fig. 2. Confusion matrices: comparison of decision tree and support vector machine (SVM) algorithms for predicting test data. 0 – negative class, 1 – positive class, horizontal axis: predicted class, vertical axis: Real class. Thus, the upper left box is the correct-negative, the lower right box the correct-positive rate.

3.3 Predicting the Weber Fracture in Medical Records

As described above, two different classification approaches were taken to predict unknown radiological findings. Figure 3 shows the prediction results exemplarily, plotted for the radiological service types.

The most often positively-predicted records were in service group "NERMRTCRC" (neurological MRT) and "NERANGIO" (neurological angiography), which implies quite bad prediction rates for the classifiers.

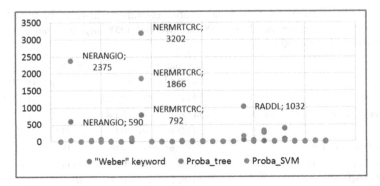

Fig. 3. Classification results: absolute occurrences of medical records predicted as "Weber"-positive according to the algorithm (Proba_tree: decision tree; Proba_SVM: support vector machine) in radiological service groups at the MHH.

3.4   Implementation into the ETL Process

We implemented the developed pipeline as a script component within the data integration task that loads the HL7 messages from the RIS. The result of the script component enriches the ECRDW core repository by splitting the findings in defined columns and by creating further information from the classification task that can be used as additional features when it comes to querying the data for Weber fracture.

4   Discussion and Conclusion

In this work the Weber fracture was to be identified in radiological medical findings. We used a dataset containing pseudonymised master data of a patient, ICD10-GM diagnosis code, as well as the free text, localisation, and certainty of every diagnosis captured during the hospital stay. The dataset also contains the radiological finding, that was split into the four fields, and additional metadata from the RIS (e.g. service group, service type text, analysis device, operating department, observation time). Various ML techniques were applied and compared with each other with regard to their suitability, accuracy and specificity. The results described in this paper show the basically feasibility of classifying texts using ML techniques. However, the results differ considerably depending on the chosen method.

The pre-selection of possible algorithms was based on algorithms that were used in the literature for similar questions: Naive Bayesian classifiers are used in many works when it comes to text classifications and sentiment analyses. The advantage of decision trees is their simple application and comprehensible interpretation [8]. SVM have been used in a number of studies, as soon as there were high dimensional vector spaces [9]. SVMs are used in a variety of problems, such as clustering, regression or classifications. The invaluable advantage of SVMs is that they work even if the features differ in test and training data sets. SVMs generate a higher dimensional vector space based on similar words by embedding the unknown tokens. Although the present paper does not use other features in the unknown data set for classification than in training, this approach could be tested in a further study.

The use of a regular expression for additional filtering of date and time information was not sensitive enough. As a result, dimensions were created in both the training and prediction datasets that might have been superfluous. Nevertheless, the use of a regular expression tokenizer showed a significant reduction of the dimensions. In a direct comparison of the classifiers, in which the dimensions were created with or without RegexpTokenizers, a clear change up to doubling of the accuracy was shown in some individual experiments.

In addition, an overfitting of the models must be considered: the training data set was created from findings texts of the radiology department of the MHH. Accordingly, it can be assumed that the models cannot easily be applied to applications at other universities or even other departments.

Classifications of medical texts, such as findings or doctor's letters, have become increasingly important in recent years, especially since they are increasingly digital and therefore available in machine-readable form [10].

No synonyms were considered in selecting the training data. Thus, we expected to have a positive prediction by only having the token "Weber" within the data. Surprisingly, this was not the case: records without "Weber" were also classified positively and vice versa. Based on the occurrences in the full-text search, we expect the SVM algorithm to be more accurate in its prediction, which is quite the opposite of what we expected from the 10-fold cross-validation and confusion matrix. To have this hypothesis confirmed, a domain expert has to validate the results, which we will do so in our next steps. Additionally, further work will include a self-generating dictionary of synonyms by implementing word embeddings to increase the recall.

Another promising approach would be to use a semi-supervised learning method: first training would be performed by using a little fraction of the whole data and would then be post-trained continuously by various prediction and validation rounds on real data. It would be promising to combine the semi-supervised techniques with a word embedding.

In summary, we have shown that implementing a NLP approach into a data warehouse ETL pipeline with Python is feasible. The developed pipeline provides more flexibility for data pre-processing and data cleansing of unstructured or semi-structured information than we would have had by using the standard data integration services of MS SQL Server. Additionally, adding a data mining pipeline for a specific research question upon this data is applicable, but its power definitively relies on validated gold standard training data and the validation of the predictions provided by a clinical expert, which will be our next step. Further limitations are due to the chosen dataset: As a medical record may have more than one medical diagnosis (e.g. differences in entry diagnosis and in release diagnosis), it may be rated as well as positive or as negative.

Nevertheless, this study proves the possibility of combining ETL processes with machine learning techniques. For one's own attempt of implementing, the pipeline has to be adapted to one's own IT infrastructure, since every hospital has its own, heterogeneous infrastructure and conditions.

References

1. Köppen, V., Saake, G., Sattler, K.-U.: Data Warehouse Technologien. MITP (2014). ISBN 9783826694851
2. Tolxdorff, T., Puppe, F.: Klinisches Data Warehouse. Informatik-Spektrum **39**, 233–237 (2016). https://doi.org/10.1007/s00287-016-0968-3
3. Zapletal, E., Bibault, J.-E., Giraud, P., Burgun, A.: Integrating multimodal radiation therapy data into i2b2. Appl. Clin. Inform. **09**, 377–390 (2018). https://doi.org/10.1055/s-0038-1651497
4. Dietrich, G., et al.: Ad hoc information extraction for clinical data warehouses. Methods Inf. Med. **57**, e22–e29 (2018). https://doi.org/10.3414/ME17-02-0010
5. Kharat, A., Singh, A., Kulkarni, V., Shah, D.: Data mining in radiology. Indian J. Radiol. Imaging **24**, 97 (2014). https://doi.org/10.4103/0971-3026.134367
6. Do, B.H., Wu, A.S., Maley, J., Biswal, S.: Automatic retrieval of bone fracture knowledge using natural language processing. J. Digit. Imaging **26**, 709–713 (2013). https://doi.org/10.1007/s10278-012-9531-1
7. Perkins, J.: Python Text Processing with NLTK 2.0 Cookbook. Packt Publishing, Birmingham (2010). ISBN 978-1-849513-60-9
8. Daumke, P., Simon, K., Paetzold, J., Marwede, D., Kotter, E.: Data-Mining in radiologischen Befundtexten. RöFo - Fortschritte auf dem Gebiet der Röntgenstrahlen und der Bildgeb. Verfahren **182**, WS117_3 (2010). https://doi.org/10.1055/s-0030-1252462
9. Kavuluru, R., Rios, A., Lu, Y.: An empirical evaluation of supervised learning approaches in assigning diagnosis codes to electronic medical records. Artif. Intell. Med. **65**, 155–166 (2015). https://doi.org/10.1016/j.artmed.2015.04.007
10. McNutt, T.R., Moore, K.L., Quon, H.: Needs and challenges for big data in radiation oncology. Int. J. Radiat. Oncol. Biol. Phys. **95**, 909–915 (2016). https://doi.org/10.1016/j.ijrobp.2015.11.032

Data Exploration in the Life Sciences

Towards Research Infrastructures that Curate Scientific Information: A Use Case in Life Sciences

Markus Stocker[1,2](✉)[iD], Manuel Prinz[1][iD], Fatemeh Rostami[3][iD],
and Tibor Kempf[3]

[1] TIB Leibniz Information Centre for Science and Technology,
Welfengarten 1 B, 30167 Hannover, Germany
{markus.stocker,manuel.prinz}@tib.eu
[2] PANGAEA Data Publisher for Earth & Environmental Science,
MARUM Center for Marine Environmental Sciences,
Leobener Strasse 8, 28359 Bremen, Germany
mstocker@marum.de
[3] Division of Molecular and Translational Cardiology,
Department of Cardiology and Angiology, Hannover Medical School,
Carl-Neuberg-Strasse 1, 30625 Hannover, Germany
{rostami.fatemeh,kempf.tibor}@mh-hannover.de

Abstract. Scientific information communicated in scholarly literature remains largely inaccessible to machines. The global scientific knowledge base is little more than a collection of (digital) documents. The main reason is in the fact that the document is the principal form of communication and—since underlying data, software and other materials mostly remain unpublished—the fact that the scholarly article is, essentially, the only form used to communicate scientific information. Based on a use case in life sciences, we argue that virtual research environments and semantic technologies are transforming the capability of research infrastructures to systematically acquire and curate machine readable scientific information communicated in scholarly literature.

Keywords: Scientific information
Scholarly communication · Knowledge representation
Virtual research environments · Research infrastructures
Knowledge infrastructures

1 Introduction

The critique is not new and the quest remains: Despite advances in information technology and systems, the format of the scholarly article has largely remained unchanged [16,17,32]. The wealth of scientific information conveyed by the steadily increasing number of published articles [9,27,43] continues to be confined to the document, seemingly inseparable from the medium as hieroglyphs carved in stone.

© Springer Nature Switzerland AG 2019
S. Auer and M.-E. Vidal (Eds.): DILS 2018, LNBI 11371, pp. 61–74, 2019.
https://doi.org/10.1007/978-3-030-06016-9_6

Document centric scholarly communication has its challenges. Most obviously, machine processing of the information communicated in scholarly articles is very limited. While words can be indexed and searched, the semantics of numbers, text, figures, symbols, etc. are hardly accessible to computers and modern exploration, retrieval, question answering and visualization thus not applicable. Such limited machine support hinders the efficient processing of literature since relevant information is "buried" in documents and finding information relies on sifting through documents. Given the growing scientific output, processing literature ties up increasing resources.

To be sure, important advances have been made. The interlinking of articles with related entities is a notable recent development. Aided by interoperable information infrastructures—such as DataCite, Crossref, literature and data publishers—articles are increasingly linked to related persistently identified datasets, audio/video, samples, instruments, software, people, institutions. The Scholix framework for scholarly link exchange [10] is a project that focuses on interoperability of information about the links between scholarly literature and data. Related advancements can be noticed also in systems that are well-known to researchers. Taking the link between articles and citations as an example, ResearchGate now shows citations "in context" by pointing readers directly to the relevant position in articles. Other related projects include Research Graph [3], RMap [25], and Research Objects [8]. The resulting graphs enable new forms of information publication, search, navigation and discovery. However, it is not scientific information communicated in scholarly literature that these graphs capture but information (i.e., metadata) about the digital objects used in communication and their relationships to contextual entities.

Another notable development is in technologies and vocabularies for machine readable representations of scientific information authors communicate in scholarly literature. Indeed, representing scientific knowledge claims has been explored for at least a decade. With the HypER approach, de Waard et al. [18] proposed to extract knowledge from articles "to allow the construction of a system where a specific scientific claim is connected, through trails of meaningful relationships, to experimental evidence." García-Castro et al. [22] proposed an extension to the Annotation Ontology [15] that enables the modelling of concepts and relations of scholarly articles, such as 'claim', 'hypothesis' or 'contradicts' and 'proves'. Nanopublications [23,31] is a concept and model designed to represent, in machine readable form, scientific statements. The OBO Foundry [34] publishes ontologies that include numerous relevant concepts e.g., for the machine readable representation of statistical hypothesis tests or average values. As a result, it is now possible to describe scientific information authors communicate in scholarly literature in machine readable form and thus have infrastructures curate, process, and publish such information as distinct information objects.

A third important advancement is in virtual research environments (VREs) [2,12] (also known as virtual laboratories and science gateways) that enable the execution of data analysis on interoperable infrastructure. Since VREs can be extended in functionality and engineered to meet advanced requirements, the p-value resulting in a statistical hypothesis test is no longer a mere number (as is generally the case in local computational environments) but can be an information object relating the p-value to the kind of statistical test performed, the involved continuous variables and values, and even data provenance in laboratory experiments. In other words a machine readable description of the performed statistical hypothesis test.

Based on a use case in life sciences, we argue that key technologies needed for research infrastructures to acquire and curate more of the scientific information communicated in scholarly literature as machine readable interlinked yet distinct information objects are in place. While certainly challenging, technological integration seems to be on the horizon. Here, we depict such an integration in the context of an open project[1] recently initiated by the TIB Leibniz Information Centre for Science and Technology which aims to develop infrastructure that acquires, curates, and processes scientific information communicated in scholarly literature [5]. In addition to technical considerations elucidated on the use case, we discuss possible pathways through which machine readable scientific information may be systematically acquired by the prospective infrastructure. We also present recent developments and some near-future plans of the project.

2   Use Case

We aim to reproduce and represent, in machine readable form, the statistical hypothesis test supporting the scientific statement that "IRE binding activity was significantly reduced in failing hearts" as published by Haddad et al. [24, p. 364] in their article entitled *Iron-regulatory proteins secure iron availability in cardiomyocytes to prevent heart failure* recently published by European Heart Journal.

Iron-responsive elements (IREs) are conserved nucleotide sequences located in uncoded regions of iron-related transcripts (mRNA). These elements can be bound by iron-regulatory proteins (IRPs) in order to regulate the iron homeostasis in cells, which is essential for cell survival since iron is a key co-factor for many enzymes involved in numerous biological processes, ranging from DNA synthesis to energy metabolism. In iron-depleted cells, IRP activity increases in order to secure the iron availability [26]. According to Haddad et al., patients with heart failure (a condition whereby the heart is unable to pump sufficiently) show reduced IRP activity and iron content in heart cells, leading to impaired heart function.

The statement by Haddad et al. is based on data reported in the plot shown in their Fig. 1B, specifically for non-failing hearts (NF) and patients with failing heart (F). The data reported in the plot are themselves sourced in the electrophoretic mobility shift assay shown as image in Fig. 1B. The quantification of the image is done using ImageJ [33], an image processing and analysis software.

Given the data, Haddad et al. use Prism (GraphPad Software) to perform a Student's t-test and find the reported statistical difference ($P < 0.001$) in mean

---

[1] Open Research Knowledge Graph: http://orkg.org (Accessed: October 16, 2018).

IRE binding activity between the two groups (NF and F). Hence the author's statement that "IRE binding activity was significantly reduced in failing hearts." Prism is also used to create the plot shown in Fig. 1B.

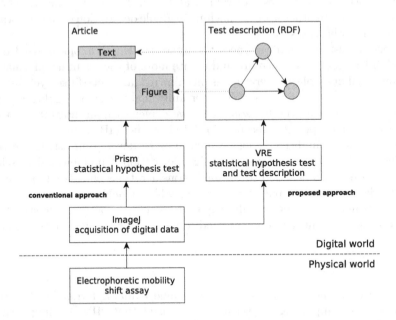

Fig. 1. Overview of the main aspects of the conventional and proposed approaches.

3   Architecture

Figure 1 contrasts the main aspects of the conventional approach just described with those of the proposed one. In the conventional approach, the proposed one adopts a system architecture with technical and social subsystems, and sociotechnical subsystem integration. However, subsystems differ in details.

In the proposed approach, the technical subsystem consists of a digital infrastructure that operates a semantically enhanced Virtual Research Environment (VRE). While VREs typically support numerous features e.g., cataloguing and communication, of primary concern here is a component for data analysis. It is this VRE component that we suggest to semantically enhance. The technical subsystem also consists of a component capable of storing and retrieving information objects. The social subsystem consists of individual researchers, members of research communities. Among other activities, researchers are the agents that perform data analysis. The proposed approach also relies on sociotechnical integration. Indeed, researchers are required to move data analysis from local computing environments into the VRE. This is to ensure that the data derived in analysis conform with the representational requirements of the system.

Data analysis is the key activity that evolves uninterpreted data to scientific information, ultimately published in scholarly literature. We borrow the notion of *data interpretation* from the unified definitional model of data, information, and knowledge proposed by Aamodt and Nygård [1]. According to the model, data are uninterpreted symbols with "no meaning for the system concerned" and are input to an interpretation process. Information is interpreted data i.e., data with meaning and the output from data interpretation. Interpretation occurs "within a real-world context and for a particular purpose." Aamodt and Nygård's model also defines knowledge as learned information. As the output of learning processes, "knowledge is information incorporated in an agent's reasoning resources."

Floridi [21] further elaborates the definition of information. Building on a widely adopted General Definition of Information (GDI), he develops a definition of semantic information. GDI defines information in terms of "data + meaning." Floridi proposes a more precise formulation that borrows the term *infon* [7,19], a discrete item of information. The infon σ is an instance of information, understood as *semantic content*, if and only if σ consists of n data, $n \geq 1$; the data are well formed; and the well-formed data are meaningful (i.e., of significance to some person, situation or machine). Of specific interest here is *factual* semantic content i.e., semantic content about a situation or fact that can be qualified as either true or false. Only semantic content that is true is informative. Thus, Floridi suggests that p qualifies as factual *semantic information* if and only if p is well-formed, meaningful, and *truthful* data. Furthermore, Floridi proposes a classification of types of data, of which two are of importance here. *Primary data* are the principal data stored, for example in a database while *derivative data* are data that "can be extracted from some data whenever the latter are used as indirect sources in search of patterns, clues or inferential evidence about things other than those directly addressed by the data themselves."

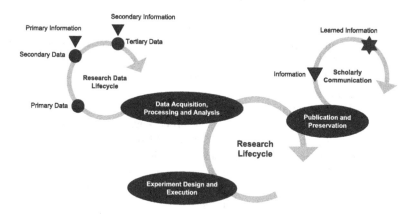

Fig. 2. From uninterpreted data to scientific information in the research lifecycle.

Listing 1. Python implementation of the statistical hypothesis test.

```python
import numpy as np
import pandas as pd
from scipy.stats import ttest_ind

labels = ['non-failing heart (NF)', 'failing heart (F)']
data = [(99, 52), (96, 40), (100, 38), (105, 18),
        (np.nan, 11), (np.nan, 5), (np.nan, 42),
        (np.nan, 55), (np.nan, 53), (np.nan, 39),
        (np.nan, 42), (np.nan, 50)]

df = pd.DataFrame.from_records(data, columns=labels)
tt = ttest_ind(df['non-failing heart (NF)'],
               df['failing heart (F)'],
               equal_var=False, nan_policy='omit')

tt.pvalue
```

Figure 2 places these concepts in the context of the research lifecycle. Uninterpreted, primary data resulting in observation, experimentation, or simulation activities enter the research data lifecycle by data acquisition. Primary data may be processed in activities other than data interpretation (e.g., aggregation or interpolation). In such activities, derived data remain uninterpreted and without meaning for the system concerned. It is in data analysis that data are interpreted and derived data are information, meaningful and—following Floridi—truthful data for the system concerned. Along research data lifecycles, data may be processed and analysed repeatedly resulting in secondary, tertiary, quaternary, etc. data and, if data are meaningful and truthful for the system concerned primary, secondary, etc. information.

Factual semantic information is a fundamental unit in scholarly communication. Figure 2 suggests that information is learned, incorporated in an agent's (researcher, primarily) "reasoning resources" (knowledge base). Through learning processes, in scholarly communication information thus evolves to knowledge, specifically learned scientific or scholarly information.

Instances of factual semantic information and learned scientific information communicated in scholarly literature are the objects which the proposed architecture aims to represent, acquire, curate, and publish for further reuse. Their representation is machine readable. Critically, not just the data that constitute information are machine readable: Meaning is machine readable, too. Hence, not only is the value 0.013 machine readable but so is its meaning as a e.g., p-value. We now present the implementation of the use case following this architecture and conceptual framework.

4   Implementation

We implement the statistical hypothesis test using Jupyter [29] in Python, specifically Jupyter Lab, the next-generation web-based interface for Project Jupyter. Jupyter Lab acts as VRE component that provides services for data analysis among the range of services typically provided by a full-fledged VRE e.g., D4Science VREs [11].

The complete Jupyter notebook is published [38]. We limit the presentation here to the key elements. Given the experimental data, the statistical hypothesis test can be easily implemented using SciPy [28]. Listing 1 shows the implementation in detail. The last line returns the computed p-value i.e., 0.0000000131. This merely reproduces in Jupyter Lab some of the output researchers obtain using Prism.

More interesting is the possibility to describe, in machine readable form, the performed statistical hypothesis test. Since our Jupyter Lab based VRE component can be extended with novel functionality, we implement a function that returns a description of the test in RDF (Resource Description Framework) [30]. Listing 2 displays the core of the description (prefixes are omitted). The numeric p-value is described as the output of a two sample t-test with unequal variance (STATO_0000304). The test description also specifies iron-responsive element binding (GO_0030350) as the study design dependent variable (OBI_0000751), a specified input of the statistical hypothesis test. Omitted here for the sake of brevity, the description also includes the continuous variables (STATO_0000251) as specified input. The input data are scalar measurement data (IAO_0000032) that are part of (BFO_0000051) the continuous variables.

Hence, rather than merely representing the numerical p-value, the approach pursued here describes the performed statistical hypothesis test in a comprehensive and semantic manner, including meaningfully described test input and output. Furthermore, the resulting description is machine readable. The description is an instance of machine readable factual semantic information communicated in scholarly literature.

Given such machine readable descriptions of statistical hypothesis tests e.g., the others included in the paper by Haddad et al. and potentially the many more found in the scientific literature, it is trivial to formulate queries only for statistically significant (specifically, $P < 0.005$ or $P < 0.001$) tests (of a specific kind) involving a particular dependent variable and continuous variables with at least N measurement data. The scientific information communicated in scholarly literature—here the statement that "IRE binding activity was significantly reduced in failing hearts," or more accurately the statistical hypothesis test underlying this statement, with the supporting figures and data in Fig. 1B— is thus not just reported in a form suitable for human experts but also available in machine readable form for automated processing.

Technically, the machine readable description of the statistical hypothesis test is a (small) RDF graph, consisting of a set of RDF triples (109 in our example). Various kinds of databases can be used to persist such triples. The most obvious kind is one of the many available triple stores. However, we are

Listing 2. Machine readable description of the performed statistical hypothesis test, in RDF Turtle syntax. For the sake of brevity, we omit prefixes but include human readable comments to guide readers through the non-semantic names of OBO Foundry ontology concepts and relations.

```
# a two sample t-test with unequal variance
[] a obo:STATO_0000304 ;
  # that has specified input
  obo:OBI_0000293 [
    # a study design dependent variable
    a obo:OBI_0000751 ,
      # specifically, iron-responsive element binding
      obo:GO_0030350
  ] ;
  # and has specified output
  obo:OBI_0000299 [
    # a p-value
    a obo:OBI_0000175 ;
    # that has value specification
    obo:OBI_0001938 [
      # a scalar value specification
      a obo:OBI_0001931 ;
      # that has specified numeric value
      obo:OBI_0001937 1.311125e-08
    ]
  ] .
```

currently experimenting with a more general purpose graph database, specifically Neo4j (neo4j.com). The primary motivation for this choice is the possibility, in Neo4j, to attach arbitrary attributes to graph nodes and edges. We plan to make use of this feature to e.g., timestamp data and support versioning.

Aligned with RDF, at the core of our data model is the statement i.e., a structure of three elements (subject, predicate, object) whereby the subject is a resource and the object is either a resource or a literal (predicate is an additional type). Statements, resources, and predicates are identified by means of an internal identifier. With RDF data, URIs are thus mapped to internal identifiers and are, in our data model, the labels of resources or predicates.

A REST API enables interaction with the graph database. Of primary focus here, the API supports the creation and lookup of resources, predicates and statements. Given the RDF triples for the machine readable description of the statistical hypothesis test (Listing 2), we thus implement the storing of triples as statements. Contrary to conventional triple stores, we first need to resolve URIs in triple subject, predicate, and object positions to internal identifiers. Hence, before a statement is stored we perform lookups and create new resources and a predicate in case the corresponding URIs cannot be found (for more detail, see [38]). Given internal identifiers for subject, predicate and resource object we then store the statement. Literal objects are unidentified values.

5   Discussion

As suggested by Mons and Velterop for their paper [31], also this paper may appear paradoxical since "it is a paper in classical format that seems to make a plea for the ending of precisely such textual classical publication." Except that this paper is no plea for the ending of classical publication. Rather, we argue that with relatively minor changes to current research infrastructures we may achieve the co-existence of classical publication with machine readable representations of (some of) the information communicated in classical publication.

We suggest that a key element is the prospective (*a priori*) systematic acquisition of machine readable scientific information communicated in scholarly literature i.e., acquisition while researchers perform data analysis and develop the results that build the foundation for the prospective article. This stands in contrast with the (complementary) approach whereby machine readable scientific information is extracted retrospectively (*a posteriori*) from published articles, principally using text mining, possibly combined with human curation.

As shown with our use case, the prospective approach has the potential to capture scientific information at the granularity of individual statements or even numbers reported in tables and figures. We argue that, with current technologies, such granularity cannot be achieved by the retrospective approach, using text mining.

However, the prospective approach relies on changes to the research infrastructure used for data analysis. The challenges are both technical and social. The technical infrastructure needs to be advanced so that the output of computational environments are no longer mere numbers. Rather, numbers need to be information objects with machine readable serialization that captures meaning. Furthermore, the technical infrastructure needs to automatically track relations between entities e.g., to record provenance.

Infrastructure is invisible [35] and this is precisely how the additional functionality delineated here should appear to researchers: invisible. However, some changes in practice are difficult to avoid. Moving data analysis from local computing environments onto interoperable infrastructure e.g., into VREs that interoperate with data and computing resources, is a major change to how data analysis is currently performed, by many if not most researchers and especially those working with little data. Data analysis on local computing environments (e.g., the researcher's workstation) is a key reason for the staggering syntactic and semantic heterogeneity of derivative data generated by researchers in data analysis. In such environments it is hard to harmonize data representation, introduce novel approaches and promote interoperability. Furthermore, the infrastructural discontinuity between local computing environments and engineered research infrastructures makes it difficult or impossible for the latter to monitor workflows and thus track executed activities, retain information about the involved primary and derivative data, as well as to systematically acquire derivative data. Indeed, the download of data from research infrastructures e.g., data repositories is "considered harmful" in most cases [4]. Implications in disciplines with sensitive, personal data such as life sciences need to be considered.

While moving data analysis onto interoperable infrastructure is surely a major social challenge for many research groups and communities, the prospect of performing more of data analysis in well-engineered VREs has great potential as an approach to start addressing the issues discussed here. Naturally, "big science" and "big data" research communities have taken steps into such direction. For example, with CERN Analysis Preservation [14] the High-Energy Physics community is systematically preserving research objects (e.g., data, software) created in analysis. However, the long tail of research with "small data" has arguably been left behind.

The proposed approach can be discussed from the perspective of the FAIR principles for scientific data management and stewardship [44]. The content of Listing 2 is of course data. As they encode scientific information communicated in scholarly literature, the data in Listing 2 are, however, of a kind different from observational data (e.g., sensor network sourced), experimental data (e.g., assay sourced) or computational data (e.g., simulation sourced). In contrast to the form in the article by Haddad et al. (in Fig. 1B and in the main text of the article) the data in Listing 2 are clearly more (machine) interoperable. Indeed, the data meet the three requirements for interoperability suggested by the FAIR principles. Specifically, in the proposed form the data are more interoperable because they "use a formal, accessible, shared, and broadly applicable language for knowledge representation"; they "use vocabularies that follow FAIR principles"; and they "include qualified references to other (meta)data." With systematic acquisition in research infrastructures, the proposed approach also supports the findability, accessibility and reusability of scientific information published in scholarly literature, and hence improves on the other elements of the FAIR principles.

The reference to concepts e.g., two sample t-test with unequal variance (STATO_0000304) and their formal semantics by means of global and unambiguous identifiers is a key aspect of the FAIR principles. In the proposed approach, infrastructure adopts identified concepts of existing ontologies. The semantics of the resulting data (Listing 2) are thus accessible to machines. This stands in contrast with the natural language text of the original study in which the authors did not make use of ontology concepts.

We implement the proposed approach in Python. With the rdflib[2] library, the language has good support for RDF. It is thus straightforward to implement the proposed features in Python. Jupyter supports numerous languages, including R which is another language popular in data science. The effort required to implement the proposed approach in Jupyter but for another language thus depends primarily on whether or not there exists a corresponding RDF library. More flexible approaches may be engineered.

---

[2] https://rdflib.readthedocs.io/ (Accessed: October 16, 2018).

Listing 2 only shows iron-responsive element binding (GO_0030350) and the p-value as statistical hypothesis test input and output, respectively. The published Jupyter notebook [38] also includes the data as specified test input. In principle, this description can be extended with further attributes. However, such extension relies on additional vocabulary, likely of a different ontology. For instance, it may be interesting for applications to explicitly capture data summaries e.g., the sample size or the share of NaN values. Such indicators are important in data and statistical test quality assessment. Furthermore, we may capture additional medical context (e.g., ICD-11 codes). To be useful, it is essential for descriptions to adequately capture context. So far, we have given this aspect only limited attention.

6   Future Work

Though some of the foundations for the infrastructure depicted here have been laid in other disciplinary contexts, specifically the earth and environmental sciences [36, 37, 39–42], the presented work remains in an embryonic stage. Most of the work required to make the vision [5, 6, 20] reality surely lays ahead. We present here a few avenues for future work.

The application of the approaches originally developed in use cases in earth and environmental sciences to life sciences is important and we are committed to build on the results reported here and develop a compelling use case together with Hannover Medical School as a research infrastructure in life sciences. Such collaboration is essential to determine the requirements for a viable infrastructure.

There exist numerous pathways along which machine readable scientific information can be acquired. In this paper, our focus is on the prospective pathway with data analysis. Also in the category of prospective pathways, we will explore the possibility of acquiring machine readable scientific information at the time of writing the article. Here, it is possible to link existing information objects created e.g., during data analysis with the article. We will explore collaboration with projects such as Dokieli [13] and other document authoring systems.

The retrospective pathways form a further category. They assume one or more written articles, extract scientific information from them, and represent information in machine readable form. In addition to text mining articles, it is interesting to explore the acquisition of machine readable scientific information at the time of article submission. This could be achieved in collaboration with submission systems, such as EasyChair. In addition to metadata about the article, such systems increasingly capture other information e.g., ORCID iDs and funding data. While it is of course untenable to expect a complete "semantification" of the article by the researcher at this point, it is arguably possible to present researchers with a form that captures the key aspects of the research contribution. Text mining could support researchers with suggestions.

As an open project, TIB encourages active stakeholder participation. The project's workshop series is a key instrument to this effect. We invite domain scientists to contribute requirements, use cases and domain expertise; representatives of related projects such as FREYA, OpenAIRE, Research Graph to explore synergies among infrastructures; representatives of the publishing sector (articles, data and other artefacts) for their related work and possible future integrations.

7   Conclusion

For a use case in life sciences, we have demonstrated how research infrastructures can systematically acquire machine readable scientific information communicated in scholarly literature. We argue that this possibility is enabled by the technological integration of VREs (in particular components for data analysis) and semantic technologies. While technical challenges do exist, we argue that the greater challenges are social, specifically the required changes in research practices. Indeed, data analysis currently performed on local computing environments needs to move into VREs. Such environments can be engineered to include novel functionality that enables the systematic acquisition of scientific information so that information is also represented in machine readable form using technologies that not only represent data but also their meanings.

Acknowledgements. We thank the TIB Leibniz Information Centre for Science and Technology for supporting this project and our colleagues and the participants of the project's workshop series for their contributions.

References

1. Aamodt, A., Nygård, M.: Different roles and mutual dependencies of data, information, and knowledge - an AI perspective on their integration. Data Knowl. Eng. **16**(3), 191–222 (1995)
2. Allan, R.: Virtual Research Environments: From Portals to Science Gateways. Chandos Publishing, Oxford (2009)
3. Aryani, A., Wang, J.: Research graph: building a distributed graph of scholarly works using research data switchboard. In: Open Repositories Conference (2017)
4. Atkinson, M., Filgueira, R., Spinuso, A., Trani, L.: Download considered harmful (2018). Manuscript in preparation
5. Auer, S.: Towards an open research knowledge graph, January 2018
6. Auer, S. Kovtun, V., Prinz, M., Kasprzik, A., Stocker, M., Vidal, M.E.: Towards a Knowledge Graph for Science. In: Proceedings of the 8th International Conference on Web Intelligence, Mining and Semantics, WIMS 2018, pp. 1:1–1:6. ACM, New York (2018)
7. Barwise, J., Perry, J.: Situations and attitudes. J. Philos. **78**(11), 668–691 (1981)
8. Bechhofer, S., Roure, D.D., Gamble, M., Goble, C., Buchan, I.: Research objects: towards exchange and reuse of digital knowledge. In: Nature Precedings, July 2010
9. Bornmann, L., Mutz, R.: Growth rates of modern science: a bibliometric analysis based on the number of publications and cited references. J. Assoc. Inf. Sci. Technol. **66**(11), 2215–2222 (2015)

10. Burton, A.: The Scholix framework for interoperability in data-literature informa-
    tion exchange. D-Lib Mag. **23**(1/2) (2017)
11. Candela, L., Castelli, D., Pagano, P.: D4Science: an e-infrastructure for support-
    ing virtual research environments. In: Agosti, M., Esposito, F., Thanos, C. (eds)
    Proceedings of the 5th Italian Research Conference on Digital Libraries (IRCDL
    2009), Padova January 2009
12. Candela, L., Castelli, D., Pagano, P.: Virtual research environments: an overview
    and a research agenda. Data Sci. J. **12** GRDI75-GRDI81 (2013)
13. Capadisli, S., Guy, A., Verborgh, R., Lange, C., Auer, S., Berners-Lee, T.: Decen-
    tralised authoring, annotations and notifications for a read-write web with dokieli.
    In: Cabot, J., De Virgilio, R., Torlone, R. (eds.) ICWE 2017. LNCS, vol. 10360, pp.
    469–481. Springer, Cham (2017). https://doi.org/10.1007/978-3-319-60131-1_33
14. Chen, X., Dallmeier-Tiessen, S., Dani, A., Dasler, R., Fernández, J.D., Fokianos,
    P., Herterich, P., Šimko, T.: CERN analysis preservation: a novel digital library
    service to enable reusable and reproducible research. In: Fuhr, N., Kovács, L., Risse,
    T., Nejdl, W. (eds.) Research and Advanced Technology for Digital Libraries. pp,
    pp. 347–356. Springer International Publishing, Cham (2016)
15. Ciccarese, P., Ocana, M., Castro, L.J.G., Das, S., Clark, T.: An open annotation
    ontology for science on web 3.0. J. Biomed. Semant. **2**(2), S4 (2011)
16. de Sompel, H.V., Payette, S., Erickson, J., Lagoze, C., Warner, S.: Rethinking
    scholarly communication. D-Lib Mag. 10(9), (2004)
17. de Waard, A., Breure, L., Kircz, J.G., van Oostendorp, H.: Modeling rhetoric in
    scientific publications. In Proceedings of the International Conference on Multidis-
    ciplinary Information Sciences and Technologies (InSciT 2006) (2006)
18. de Waard, A., Shum, S.M., Carusi, A., Park, J., Samwald, M., Sándor, Á.: Hypothe-
    ses, evidence and relationships: the HypER approach for representing scientific
    knowledge claims. In: Clark, T., Luciano, J.S., Marshall, M.S., Prud'hommeaux,
    E.., Stephens, S. (eds), Proceedings of the Workshop on Semantic Web Applica-
    tions in Scientific Discourse (SWASD 2009), vol. 523, Washington October 2009.
    CEUR
19. Devlin, K.: Logic and Information. Cambridge University Press, Cambridge (1991)
20. Fathalla, S., Vahdati, S., Auer, S., Lange, C.: Towards a knowledge graph repre-
    senting research findings by semantifying survey articles. In: Kamps, J., Tsakonas,
    G., Manolopoulos, Y., Iliadis, L., Karydis, I. (eds.) Research and Advanced Tech-
    nology for Digital Libraries. pp, pp. 315–327. Springer International Publishing,
    Cham (2017). https://doi.org/10.1007/978-3-319-67008-9_25
21. Floridi, L.: The Philosophy of Information. Oxford University Press, Oxford (2011)
22. García-Castro, L.J., Giraldo, O.X., García-Castro, A.: Using annotations to model
    discourse: an extension to the annotation ontology. In García-Castro, A. Lange, C.,
    van Harmelen, F., Good, B. (eds), Proceedings of the 2nd Workshop on Semantic
    Publishing, vol. 903, pp. 13–22, Hersonissos, May 2012. CEUR
23. Groth, P., Gibson, A., Velterop, J.: The anatomy of a nanopublication. Inf. Serv.
    Use **30**(1–2), 51–56 (2010)
24. Haddad, S.: Iron-regulatory proteins secure iron availability in cardiomyocytes to
    prevent heart failure. Eur. Heart J. **38**(5), 362–372 (2017)
25. Hanson, K.L., DiLauro, T., Donoghue, M.: The RMap project: capturing and pre-
    serving associations amongst multi-part distributed publications. In: Proceedings
    of the 15th ACM/IEEE-CS Joint Conference on Digital Libraries, JCDL 2015, pp.
    281–282. ACM, New York (2015)
26. Hentze, M.W., Muckenthaler, M.U., Galy, B., Camaschella, C.: Two to tango:
    regulation of mammalian iron metabolism. Cell **142**(1), 24–38 (2010)

27. Jinha, A.E.: Article 50 million: an estimate of the number of scholarly articles in existence. Learn. Publishing **23**(3), 258–263 (2010)

28. Jones, E., Oliphant, T., Peterson, P. et al.: SciPy: Open source scientific tools for Python (2001)

29. Kluyver, T.: Jupyter notebooks–a publishing format for reproducible computational workflows. In: Loizides, F., Schmidt, B. (eds), Positioning and Power in Academic Publishing: Players, Agents and Agendas, pp. 87–90. IOS Press (2016)

30. Manola, F., Miller, E., McBride, B.: RDF Primer. W3C Recommendation **10**(1–107), 6 (2004)

31. Mons, B., Velterop, J.: Nano-publication in the e-science era. In: Workshop on Semantic Web Applications in Scientific Discourse (SWASD 2009), Washington (2009)

32. Priem, J.: Beyond the paper. Nature **495**(7442), 437–440 (2013)

33. Schneider, C.A., Rasband, W.S., Eliceiri, K.W.: NIH Image to imageJ: 25 years of image analysis. Nat. Methods **9**(7), 671–675 (2012)

34. Smith, B., et al.: The OBO foundry: coordinated evolution of ontologies to support biomedical data integration. Nat. Biotechnol. **25**(11), 1251–1255 (2007)

35. Star, S.L.: The ethnography of infrastructure. Am. Behav. Sci. **43**(3), 377–391 (1999)

36. Stocker, M.: Advancing the software systems of environmental knowledge infrastructures. In: Chabbi, A., Loescher, H.W. (eds.) Terrestrial Ecosystem Research Infrastructures: Challenges and Opportunities, pp. 399–423. CRC Press, Taylor & Francis Group (2017)

37. Stocker, M.: From data to machine readable information aggregated in research objects. D-Lib Mag. **23**(1/2) (2017)

38. Stocker, M.: Jupyter notebook for DILS 2018 paper on research infrastructures that curate scientific information. Figshare, July 2018

39. Stocker, M., Baranizadeh, E., Portin, H., Komppula, M., Rönkkö, M., Hamed, A., Virtanen, A., Lehtinen, K., Laaksonen, A., Kolehmainen, M.: Representing situational knowledge acquired from sensor data for atmospheric phenomena. Environ. Model. Softw. **58**, 27–47 (2014)

40. Stocker, M., et al.: Representing situational knowledge for disease outbreaks in agriculture. J. Agric. Inf. **7**(2), 29–39 (2016)

41. Stocker, M., Paasonen, P., Fiebig, M., Zaidan, M.A., Hardisty, A.: Curating scientific information in knowledge infrastructures. Data Sci. J. **17** (2018). https://doi.org/10.5334/dsj-2018-021

42. Stocker, M., Rönkkö, M., Kolehmainen, M.: Situational knowledge representation for traffic observed by a pavement vibration sensor network. IEEE Trans. Intell. Transp. Syst. **15**(4), 1441–1450 (2014)

43. White, K.E., Robbins, C., Khan, B., Freyman, C.: Science and engineering publication output trends: 2014 shows rise of developing country output while developed countries dominate highly cited publications. Technical Report NSF 18–300, National Science Foundation, October 2017

44. Wilkinson, M.D., et al.. The FAIR guiding principles for scientific data management and stewardship. Scientific Data, 3 March 2016

Interactive Visualization for Large-Scale Multi-factorial Research Designs

Andreas Friedrich[1,2](✉) (iD), Luis de la Garza[1] (iD), Oliver Kohlbacher[1,2,3,4] (iD),
and Sven Nahnsen[1] (iD)

[1] Quantitative Biology Center (QBiC), University of Tübingen,
Auf der Morgenstelle 10, 72076 Tübingen, Germany
[2] Center for Bioinformatics, Department of Computer Science,
University of Tübingen, Sand 14, 72076 Tübingen, Germany
andreas.friedrich@uni-tuebingen.de
[3] Biomolecular Interactions, Max Planck Institute for Developmental Biology,
Max-Planck-Ring 5, 72076 Tübingen, Germany
[4] Institute for Translational Bioinformatics, University Hospital Tübingen,
Tübingen, Germany

Abstract. Recent publications have shown that the majority of studies cannot be adequately reproduced. The underlying causes seem to be diverse. Usage of the wrong statistical tools can lead to the reporting of dubious correlations as significant results. Missing information from lab protocols or other metadata can make verification impossible. Especially with the advent of Big Data in the life sciences and the hereby-involved measurement of thousands of multi-omics samples, researchers depend more than ever on adequate metadata annotation. In recent years, the scientific community has created multiple experimental design standards, which try to define the minimum information necessary to make experiments reproducible. Tools help with creation or analysis of this abundance of metadata, but are often still based on spreadsheet formats and lack intuitive visualizations. We present an interactive graph visualization tailored to experiments using a factorial experimental design. Our solution summarizes sample sources and extracted samples based on similarity of independent variables, enabling a quick grasp of the scientific question at the core of the experiment even for large studies. We support the ISA-Tab standard, enabling visualization of diverse omics experiments. As part of our platform for data-driven biomedical research, our implementation offers additional features to detect the status of data generation and more.

Keywords: Experimental design · Aggregation graph · Metadata
Portal · Reproducibility

1 Introduction

The reproducibility crisis has revealed obvious shortcomings of modern biomedical experimental techniques. While outright fraud seems to be the exception,

© Springer Nature Switzerland AG 2019
S. Auer and M.-E. Vidal (Eds.): DILS 2018, LNBI 11371, pp. 75–84, 2019.
https://doi.org/10.1007/978-3-030-06016-9_7

recent publications pinpoint many of the problems as based on missing statistical understanding when planning or performing scientific studies [4]. Even if enough data is available to draw significant conclusions, the interaction between different variables has to be reflected in the experimental design. Independent variables are usually the focus of a study and are controlled by the experimenter. However, it is rare that a variable like a disease state depends on only one single variable: regulatory networks commonly include proteins that act together to create a phenotype [12]. It is thus sensible to study multiple independent variables in a factorial experimental design. The concept of factorial experimental designs was popularized in crop research [5,6] and allows experimenters to detect interactions, something not possible in one-factor-at-a-time (OFAAT) experiments.

The advancement of Big Data assists to conduct sophisticated experiments benefiting from these study designs. Yet, even well-designed studies can often not be reproduced, because crucial metadata is missing [23]. Convenient interfaces between experimenters' notes and online database systems are often missing. Excel spreadsheets are still the most widely used tool for research notes pertaining to assays and samples [16]: early efforts to standardize scientific reporting lead to formalized spreadsheet formats specifying the minimum required information to reproduce an experiment. MIAME, the Minimum Information About a Microarray Experiment [2,3] standard and the microarray gene expression markup language MAGE-ML [20] aim at annotating experiments so they can be independently verified. Similarly, MIAPE, a standard describing the Minimum Information About a Proteomics Experiment, tries to specify the needed information to interpret analyses performed on proteins [21]. ISA-Tab combines these earlier approaches into an interoperable spreadsheet format relating information about research aims, other related studies and their associated assays [18,19]. Different efforts have been undertaken to provide users with tools based on the ISA standard [9,10]. linkedISA leverages the data provided to create a semantic, interoperable presentation and shows how implicitly defined study groups can be extracted from ISA-Tab. These groups are summarized and listed in Bio-GraphIIn, a graph-based repository for biological experimental data [8]. With the growing complexity of biological experiments and especially the communication thereof, efficient visualization are indispensable. However, most of the work has been focused on connecting experiments to ontology frameworks and making it machine-readable. While Bio-GraphIIn presents a list of study groups, this type of presentation can become difficult to grasp for huge experiments involving many experimental factors and other metadata. More information can only be obtained by displaying huge tables of samples.

The need to use computer-aided experimental design for large studies was previously discussed in early factorial design approaches in behavioral research, such as the online tool WEXTOR [17]. Here, the combination of every possible factor level pertaining to participants can be used to create specific web-pages that guide the corresponding subjects to their questions or tests. In high-throughput biomedical science, analysis tools that make use of experimental design information are often limited to custom formats: MaxQuant allows users to edit an Experimental Design template to relate files with sample fractions [22].

Here, we build on our intuitive interface for experiment creation leveraging proven experimental design concepts like full-factorial study design [7]. To connect experimental designs with data integration, we provide an interactive visualization tool that can summarize complex study designs based on involved species, tissues, analytes and experimental factors into an intuitive experiment graph. In an effort to comply with existing standards while allowing easy options to manage high-throughput experiments, we provide interoperability with the ISA-Tab format, and suggest a format for simplified experiment creation. The highly modular structure makes our tool a good starting point for further developments in the area of quality control and statistical power estimation.

2 Methods

2.1 Factorial Experimental Designs

In a factorial design the influences of all independent experimental variables on the response are investigated. A factor of an experimental design is defined as one such variable that is being studied. A level is one possible variation of a factor. The number of levels denotes the total number of different variations for a single factor that was used in an experiment. Factorial designs are called full-factorial designs, if every possible combination of levels is tested. A full-factorial experiment with n factors and k levels for each factor is called a $k \times n$ factorial design and consists of k^n sub-experiments, as exemplified in Table 1. Each of these cases can then have multiple biological or technical replicates.

Table 1. Example of a 3×2 full-factorial experimental design. Two variables x_1 and x_2 containing three levels each are tested, leading to nine different experiments.

Variables	Experiment no.								
	1	2	3	4	5	6	7	8	9
x_1	−	−	−	+	+	+	0	0	0
x_2	−	+	0	−	+	0	−	+	0

2.2 Aggregation Graph

The hierarchical way in which omics experiments are typically performed leads to an intuitive sample graph connecting patient/model organism entities to those denoting tissue/cell extract and measured analyte entities as previously described [7,14]. The example in Fig. 1 visualizes experiments on six mice. In each case a liver sample was taken and proteins prepared for mass spectrometry analysis.

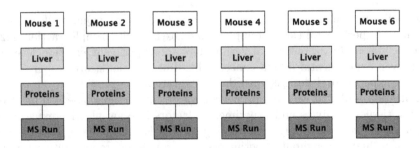

Fig. 1. Sample hierarchy of six sources with attached samples and measurements.

Let $G = (V_G, E_G)$ be a sample graph with vertices $v \in V_G$ denoting each of these entities in an experiment and edges $(v, w) \in E_G$ denoting the extraction of entity w from entity v in an experimental step.

Let further $f_1 \ldots f_n$ be a set of experimental factors on a subset of these entities with factor level f_{iv} for factor f_i of vertex v and a similarity function on factor levels $s(f_{iv}, f_{iw}) = \{0, 1\}$.

We define a set of aggregation graphs $H_1 \ldots H_n$, one for each factor f_i:

$$H_i = (V_H, E_H)$$

$$\forall\, v \in V_G : \sum_{w \in V_H} s(f_{iv}, f_{iw}) = 0 \rightarrow v \in V_H \tag{1}$$

$$\forall\, (v, w) \in E_G : v \in V_H \wedge w \in V_H \rightarrow (v, w) \in E_H$$

Each graph H aggregates all entities of G with a similar factor level into a single vertex, while preserving connections between the hierarchy levels of the experiment. For nominal factors, similarity is best defined as the perfect match of both levels, while quantitative variables can be summarized using intervals.

2.3 Implementation

We use the Open Source Biology Information System (openBIS) to store datasets and annotating metadata. Our experimental design is represented both by interconnected entities denoting source organisms and samples as well as metadata properties of these entities. Experimental factors and other properties of sample source entities and samples are stored in a intermediary XML format in openBIS that is validated by an XML schema. This schema includes quantitative variables with or without units as well as variables on a nominal scale, for example different disease states. Metadata is read from and written to the system using a web portlet running on a Liferay portal.

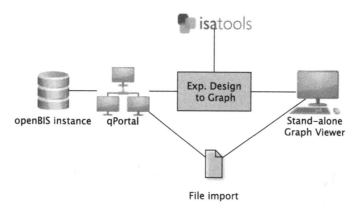

Fig. 2. Schematic diagram of our implementation: Existing experimental designs and their metadata stored in openBIS can be visualized in qPortal using our Java-based experimental design libraries. Users can import and view experiments using different formats, ISA-Tab being supported through isatools. A JavaFX implementation independent of portal or data source is available on Github.

The schematic integration of our experimental design visualization into the portal can be seen in Fig. 2. Users can create experiments using a wizard process or file import and browse information about existing experiments [7,14]. Both imported and existing experiments can be translated into the aggregation graph and displayed using Javascript libraries [1,15]. For existing experiments, meta information about attached datasets is leveraged from the data store. When importing ISA-Tab investigations, the open-source framework isatools is used in the translation process, using source and sample identifiers of the ISA study format as well as all defined experimental factors. The Javascript libraries dagre and Data-Driven Documents (D^3) are then used to compute graph coordinates and draw the the selected graph. A stand-alone version implemented in JavaFX can be used independently of the portal or openBIS.

3   Results

To compare our implementation to the usual, complete sample graph of a project, we demonstrate both visualizations on a simple proteomics experiment. 24 mice were anesthetized for different periods of time, liver tissue was extracted and proteins from those tissue samples were measured using mass spectrometry. Figure 1 shows a subset of the full sample graph. In contrast, the summarized experimental design graph of the same experiment seen in Fig. 3 gives a quick, condensed overview of the experiment hierarchy, if no factor is chosen. Metadata like sample identifiers can be shown by clicking on nodes of the graph. When the factor *anaesthesia_duration* is selected, our algorithm splits the graph into three groups of mice and descendant samples according to the three levels of this factor. Colors

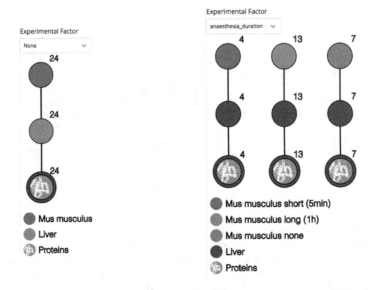

Fig. 3. Experimental design graphs of the same experiment as seen in qPortal. Numbers denote the amount of summarized samples for each factor and hierarchy level. Left: no experimental factor chosen. Right: nodes denoting mice and child nodes in the graph are split by experimental factor *anesthesia duration*.

and legend are entirely dependent on the graph and inform users about species, tissues, analytes and different factor values. Furthermore, the green outline of the protein nodes show that data generation has been completed for all samples.

We evaluate our stand-alone implementation using a recent lipidomics study on the progression to islet autoimmunity and type 1 diabetes taken from the MetaboLights database for metabolomics experiments [11,13]. Our application shows a description of the imported ISA-study and lists every experimental factor that the authors have annotated in a drop-down menu. Selecting *disease status* as seen in Fig. 4 shows that data generated from the blood plasma samples of 40 patients of a control group were compared to those of 40 type 1 diabetes (T1D) cases, as well as 40 cases of autoimmunity against islet cells, that had not yet progressed to diabetes. Selecting the *age* factor reveals that this is a time series study, where blood was taken at different ages of patients. In this case the authors failed to include units in their metadata, so it is only clear from their publication that the ages are measured in months (Fig. 5).

Our implementation is not limited to single-omics experiments. Figure 6 shows a complex multi-omics study imported via the ISA-tab format. Since the experimental factor levels only differ between cell cultures, the graph stays connected on the species level. For studies of this complexity, zoom functionality can be used to show details.

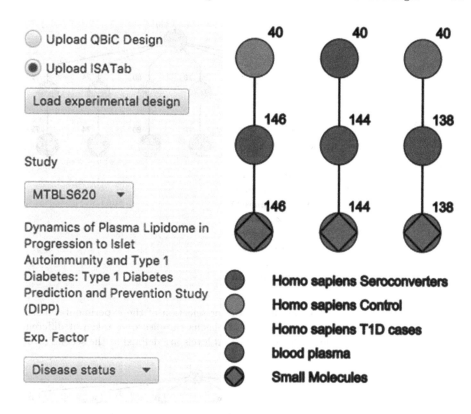

Fig. 4. View of the stand-alone application after importing an ISA-Tab folder and selecting a study as well as the experimental factor *disease status*. Information about the selected study is displayed. Users can change experimental factors from a drop-down menu. Our aggregation graph shows extraction of blood plasma from 120 patients belonging to the three groups control, type 1 diabetes (T1D) and seropositivity (for islet cell autoantibodies). The metabolome of those samples was measured.

3.1 Availability and License

The study aggregation graph is available through qPortal:
https://portal.qbic.uni-tuebingen.de/portal/web/qbic/software
A stand-alone JavaFX implementation and example studies are available on Github under the MIT license:
https://github.com/qbicsoftware/experiment-graph-gui
ISA-Tab files of the type 1 diabetes study are available on MetaboLights:
https://www.ebi.ac.uk/metabolights/MTBLS620

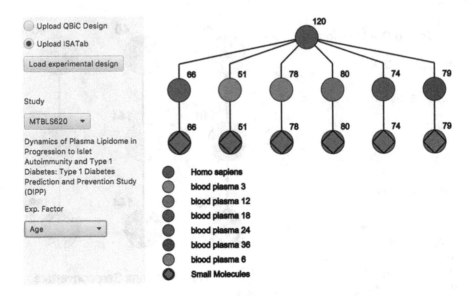

Fig. 5. View of the stand-alone application after selection of the experimental factor *age*. The levels of this factor show that blood plasma samples were taken at different ages of the same patients, since the experimental levels are defined at the second level.

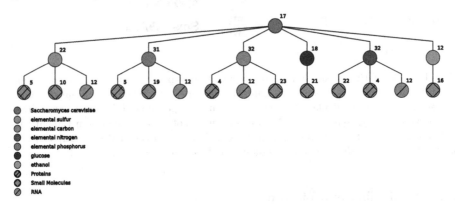

Fig. 6. Aggregation graph of one study of an imported ISA-tab investigation. Yeast cultures are grown lacking different nutrients and proteome, transcriptome and metabolome are measured.

4   Discussion

We present tools to visualize large biomedical studies by their most important experimental aspects. Building on our graphical interface for the creation of factorial experimental designs and our hierarchical data model, we create graphs summarizing complex hierarchies of experimental variables, allowing users to quickly familiarize themselves with the important aspects of a study. When used in a platform integrating experiment data and metadata, like qPortal, additional information about datasets can be leveraged, marking missing data or the status of a project. Our aggregation graph gives a concise and intuitive overview in cases where representation of experiments was previously only possible using large tables.

The lack of statistical power and sound experimental design has lead to the so-called reproducibility crisis. Extensive work has been done to standardize metadata annotation and storage, leading to multi-omics standards like ISA-Tab, which not only stores metadata, but also provides a foundation to search, display and use these annotations. These methods are clearly required due to the size of modern biomedical experiments and their metadata: a simple overview a of study often leads to huge tables or cluttered graphs. Some approaches are examples of successful, interactive uses of study design visualization, yet they address very specific questions. Bio-GraphIIn [8] focuses on listing the replicates of each study group. By contrast, our approach supports the ISA-Tab format, provides an interactive visualization of a large number of experiments and is able to summarize replicates (with respect to one factor) into a single node to display a concise representation with which users can interact to control the displayed level of detail.

We have shown that our approach can display current studies including several hundred entities. Since ISA-Tab is not a minimum information standard, the amount of actual information beyond the sample hierarchy that can be drawn from its format depends on the annotations provided by researchers, as our example shows. We have taken the first steps towards a fully modular solution that will allow integration of our tool, enforcing standards that fit with their experimental data model.

Experimental factors are one of the most important type of study annotation, since they are at the core of the question scientists want to answer. However, our concept is not necessarily bound to the aggregation of different factor levels. Any property that can split subjects or samples in different groups, can be useful to find out more about a study. In large studies involving multiple groups, sharing information about the status of the project and data generation is often important. Provided this information is available, future work could include a time-component, displaying the history of a study.

References

1. Bostock, M., Ogievetsky, V., Heer, J.: D^3 data-driven documents. IEEE Trans. Vis. Comput. Graph. **17**(12), 2301–2309 (2011)
2. Brazma, A.: Minimum information about a microarray experiment (MIAME)-successes, failures, challenges. Sci. World J. **9**, 420–423 (2009)
3. Brazma, A., et al.: Minimum information about a microarray experiment (MIAME)-toward standards for microarray data. Nat. Genet. **29**(4), 365–371 (2001)
4. Collins, F.S., Tabak, L.A.: NIH plans to enhance reproducibility. Nature **505**(7485), 612 (2014)

5. Fisher, R.: Introduction to "the arrangement of field experiments". J. Min. Agric. Gr. Br. **33**, 503–513 (1926)
6. Fisher, R.A.: The Design of Experiments. Oliver and Boyd, Edinburgh, London (1937)
7. Friedrich, A., Kenar, E., Kohlbacher, O., Nahnsen, S.: Intuitive web-based experimental design for high-throughput biomedical data. BioMed Res. Int. **2015** (2015). Article ID 958302, 8 p.
8. Gonzalez-Beltran, A., Maguire, E., Georgiou, P., Sansone, S.A., Rocca-Serra, P.: Bio-GraphIIn: a graph-based, integrative and semantically-enabled repository for life science experimental data. EMBnet. J. **19**(B), 46 (2013)
9. González-Beltrán, A., Maguire, E., Rocca-Serra, P., Sansone, S.A.: The open source ISA software suite and its international user community: knowledge management of experimental data. EMBnet. J. **18**(B), 35 (2012)
10. González-Beltrán, A., Maguire, E., Sansone, S.A., Rocca-Serra, P.: linkedISA: semantic representation of ISA-Tab experimental metadata. BMC Bioinform. **15**(14), S4 (2014)
11. Haug, K., et al.: Metabolights-an open-access general-purpose repository for metabolomics studies and associated meta-data. Nucleic Acids Res. **41**(D1), D781–D786 (2012)
12. Kanehisa, M., Goto, S., Furumichi, M., Tanabe, M., Hirakawa, M.: KEGG for representation and analysis of molecular networks involving diseases and drugs. Nucleic Acids Res. **38**(Suppl. 1), D355–D360 (2009)
13. Lamichhane, S., et al.: Dynamics of plasma lipidome in progression to islet autoimmunity and type 1 diabetes: type 1 diabetes prediction and prevention study (DIPP). bioRxiv, p. 294033 (2018)
14. Mohr, C., et al.: qPortal: a platform for data-driven biomedical research. PloS One **13**(1), e0191603 (2018)
15. Pettitt, C.: dagre - graph layout for JavaScript (2014). https://github.com/dagrejs/dagre
16. Rayner, T.F., et al.: A simple spreadsheet-based, MIAME-supportive format for microarray data: MAGE-TAB. BMC Bioinform. **7**(1), 489 (2006)
17. Reips, U.D., Neuhaus, C.: WEXTOR: a web-based tool for generating and visualizing experimental designs and procedures. Behav. Res. Methods, Instrum., Comput. **34**(2), 234–240 (2002)
18. Sansone, S.A., et al.: The first RSBI (ISA-TAB) workshop: "can a simple format work for complex studies?". OMICS J. Integr. Biol. **12**(2), 143–149 (2008)
19. Sansone, S.A., et al.: Toward interoperable bioscience data. Nat. Genet. **44**(2), 121 (2012)
20. Spellman, P.T., et al.: Design and implementation of microarray gene expression markup language (MAGE-ML). Genome Biol. **3**(9), research0046-1 (2002)
21. Taylor, C.F., et al.: The minimum information about a proteomics experiment (MIAPE). Nat. Biotechnol. **25**(8), 887–893 (2007)
22. Tyanova, S., Mann, M., Cox, J.: MaxQuant for in-depth analysis of large SILAC datasets. In: Warscheid, B. (ed.) Stable Isotope Labeling by Amino Acids in Cell Culture (SILAC): Methods and Protocols. MMB, vol. 1188, pp. 351–364. Springer, New York (2014). https://doi.org/10.1007/978-1-4939-1142-4_24
23. Vasilevsky, N.A., et al.: On the reproducibility of science: unique identification of research resources in the biomedical literature. PeerJ **1**, e148 (2013)

FedSDM: Semantic Data Manager for Federations of RDF Datasets

Kemele M. Endris[1,2]($\boxtimes$) iD, Maria-Esther Vidal[1,2] iD, and Sören Auer[1,2] iD

[1] L3S Research Center, Hannover, Germany
{endris,vidal,auer}@L3S.de
[2] TIB Leibniz Information Centre for Science and Technology, Hannover, Germany

Abstract. Linked open data movements have been followed successfully in different domains; thus, the number of publicly available RDF datasets and linked data based applications have increased considerably during the last decade. Particularly in Life Sciences, RDF datasets are utilized to represent diverse concepts, e.g., proteins, genes, mutations, diseases, drugs, and side effects. Albeit publicly accessible, the exploration of these RDF datasets requires the understanding of their main characteristics, e.g., their vocabularies and the connections among them. To tackle these issues, we present and demonstrate FedSDM, a semantic data manager for federations of RDF datasets. Attendees will be able to explore the relationships among the RDF datasets in a federation, as well as the characteristics of the RDF classes included in each RDF dataset (https://github.com/SDM-TIB/FedSDM).

1 Introduction

As the RDF data model continues gaining popularity, publicly available RDF datasets are growing in terms of number and size [2,6]. One of the challenges emerging from this trend is how to efficiently and effectively execute queries over a set of autonomous RDF datasets, i.e., a federation of RDF datasets. RDF datasets in a federation are accessible via web services, e.g., SPARQL endpoints, and a federated query processing engine enables the execution of queries over these web services. Federated query engines are responsible of selecting the relevant sources of a query, as well as of the tasks of query planning and execution, both required to collect the data from the selected sources and to answer the query [9]. Existing federated SPARQL query engines include MULDER [5], ANAPSID [1], FedX [8], SPLENDID [7], and SemaGrow [3]. Albeit effective, a federated query engine requires user queries which should be expressed in terms of the vocabularies used in the sources of a federation, as well as respecting connections among them. Consider the SPARQL query in Listing 1, that collects the mutations of the type 'confirmed somatic variant' located in transcripts which are translated as proteins that are transporters of the drug Docetaxel. To answer this query, a federated query engine should select two data sources, IASIS-KG and DrugBank. But for someone who does not have knowledge about the RDF vocabularies of these RDF datasets, writing this query

© Springer Nature Switzerland AG 2019
S. Auer and M.-E. Vidal (Eds.): DILS 2018, LNBI 11371, pp. 85–90, 2019.
https://doi.org/10.1007/978-3-030-06016-9_8

may require a great effort. We tackle the problem of exploring RDF datasets in a federation, and present and demonstrate FedSDM, a semantic data manager for federations of RDF datasets. FedSDM relies on RDF Molecule Templates (RDF-MTs) [5], abstract representations of RDF classes in the RDF datasets of a federation, and their connections. FedSDM enables the exploration of RDF-MTs; during the demonstration, attendees will observe how an RDF-MT based analysis allows for the understanding of the concepts represented in a federation, as well as the main characteristics of a federation RDF datasets.

Listing 1.1. SPARQL Query

```
SELECT DISTINCT ?mutation ?transcript
WHERE {?mutation        rdf:type iasis:Mutation .
       ?mutation    iasis:mutation_somatic_status    'Confirmed_somatic_variant'.
       ?mutation    iasis:mutation_isLocatedIn_transcript    ?transcript .
       ?transcript  iasis:translates_as              ?protein .
       ?drug        iasis:drug_interactsWith_protein ?protein .
       ?protein     iasis:label                      ?proteinName .
       ?drug        iasis:label                      'docetaxel' .
       ?drug        iasis:externalLink               ?drug1 .
       ?drug1       drugbank:transporter             ?transporter .
       ?transporter drugbank:gene-name               ?proteinName .}
```

2   The FedSDM Architecture

The FedSDM architecture includes four basic components: Metadata Manager, Metadata Explorer, Graph Analyzer, and Federated Query Engine.

Metadata Manager: is responsible for creating and managing RDF-MTs in a federation. Given a set of RDF data sources, the metadata manager creates RDF-MTs for each data source. An RDF-MT rm is described in terms of: the RDF class of rm, cardinality, set of predicates and the cardinality of each predicate, and the links to other RDF-MTs in the federation or in the same RDF dataset. The set of predicates and the links will be used to either analyze the federation or to formulate a federated query. Intra-dataset links (i.e., links between RDF-MTs within the same data source) and inter-dataset links (i.e., links between RDF-MTs from different data sources) in a federation will be exploited by other components, such as a graph analyzer to compute the graph properties of an RDF-MT network, and the federated query engine for decomposition, source selection, and planning of a federated query. **Metadata Explorer:** uses RDF-MTs created by the metadata manager to generate different visualizations. For instance, it analyzes RDF-MT links to visualize the connectivity among datasets. In addition, it acts as a gateway to access the metadata stored for further analysis of the data sources, e.g., cardinality and predicates. **Graph Analyzer:** performs graph analysis of a graph created by using intra- and inter-dataset links among RDF-MTs. Properties such as graph density, number of connected components, transitivity, and clustering coefficient of an RDF-MT graph are generated using networkx[1] python library. **Federated Query Engine:** provides a unified view

---

[1] https://networkx.github.io/.

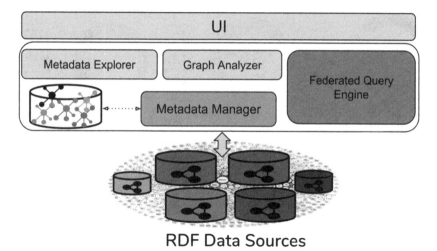

Fig. 1. FedSDM architecture. Given a set of data source endpoints by the user, the metadata manager creates the RDF-MTs from each endpoint and store it to a triple store. Metadata Explorer and Graph Analyzer issues a SPARQL query to collect basic information about RDF-MTs in the federation to perform analysis and present to user via UI component. Finally, given a SPARQL query the Federated Query Engine selects relevant data sources in the federation and execute a federated query then predent the results to the user via the UI.

of the data sources in the federation. This component exploits the metadata collected from the RDF datasets in a federation for decomposition and source selection. In FedSDM, MULDER [5] is integrated as the federated query engine (Fig. 1).

3   Demonstrating Use Cases

We created a federation composed of five data sources; DBpedia (3.5.1), Drug-Bank (Bio2RDF), PharmGKB (Bio2RDF), Sider (Bio2RDF), and the IASIS-KG. Attendees will be able to explore the RDF-MTs of these RDF datasets and their connections. Specifically, we will demonstrate the following use cases (Table 1):

Analysis of Datasets in a Federation. We will present analysis of datasets in the federation in terms of RDF-MT connectivity within a dataset and with other data sources in the federation. First, we will show the RDF-MT composition in different levels, per data source, e.g., Figs. 2a, b, c, d and e show concepts in IASIS-KG, DBpedia (3.5.1), Drugbank (Bio2RDF), Sider and PharmGKB, respectively. In addition, the federation in terms of RDF-MTs is depicted in Fig. 2f. This gives the idea on how many concepts represented in a dataset and the number of unique entities per concept. Then, we will show the connectivity among sources via RDF-MTs as a force graph (e.g., Fig. 3) and circular

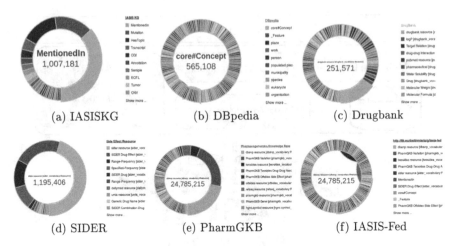

(a) IASISKG   (b) DBpedia   (c) Drugbank

(d) SIDER   (e) PharmGKB   (f) IASIS-Fed

Fig. 2. RDF-MT composition of data sources. Each colored portions represents an RDF-MT proportional to total distinct molecules in them.

Table 1. Graph metrics

Num of nodes	767
Num of edges	8698
Graph density	0.0296
Avg. num neighbors	22.6805
Connected components	67
Transitivity	0.0585
Clustering coefficient	0.0811

Fig. 3. Source links

Fig. 4. Links

(e.g., Fig. 4), demonstrating the connectivity among the RDF-MTs in the federation. Finally, we will show the graph properties, in numbers, of each data source and overall federation. Table 2 shows graph property values, such as density, connected components, transitivity, and average clustering coefficient, for each data sources and the overall federation. Average clustering coefficient assigns higher scores to low degree nodes, while the transitivity ratio places more weight on the high degree nodes.

Exploratory Metadata Analysis. In this use case, the attendee will explore the metadata of the federation to understand the characteristics of an RDF-MT, as in Fig. 5, and formulate a federated query, e.g., Fig. 6. After formulating the federated query by exploring RDF-MT properties, the query will be executed by a federated query processing engine integrated in FedSDM and results will be displayed, Fig. 7. Results can be exported as CSV, TSV, Excel, or PDF formats.

Table 2. IASIS-federation RDF-MTs graph properties. C.C - Connected Components, Avg. C - Average Clustering

Data source	Nodes	Links	Density	C.C	Avg. clus	Transitivity	Avg. neighbours
IASIS-KG	31	36	0.07741	5	0.18817	0.265822	2.3225
DBpedia	467	8124	0.07466	20	0.12097	0.05968	34.79229
DrugBank	207	353	0.01655	22	0	0	3.4106
PharmaGKB	181	273	0.01675	39	0	0	3.01657
SIDER	27	43	0.12250	5	0	0	3.18518
ALL(Fed)	767	8698	0.0296	67	0.0811	0.0585	22.6805

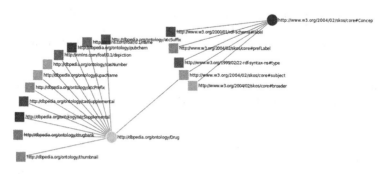

Fig. 5. `dbo:Drug` predicates & links

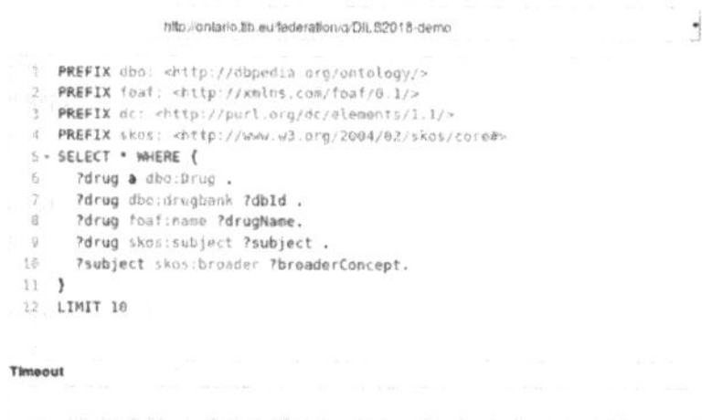

Fig. 6. SPARQL query based on the `dbo:Drug` metadata

Fig. 7. `dbo:Drug` query results (tabular)

4   Conclusions and Future Work

We present FedSDM, a semantic data manager for data federation and analysis. FedSDM provides a visual analysis of data sources and a federation by using RDF Molecule Templates. FedSDM provides an exploratory analysis on the metadata of the federation sources. In addition, FedSDM able to generate basic graph properties of RDF-MT graph. For future work, we plan to extend FedSDM to support domain specific visualization of SPARQL query results and analysis of data sources via sampling. Furthermore, FedSDM will be equipped with a component to define privacy and access control rules that must be enforced during federated query processing [4].

Acknowledgements. This work has been funded by the EU H2020 RIA under the Marie Skłodowska-Curie grant agreement No. 642795 (WDAqua) and EU H2020 Programme for the projects with GA No. 727658 (IASIS).

References

1. Acosta, M., Vidal, M.-E., Lampo, T., Castillo, J., Ruckhaus, E.: ANAPSID: An adaptive query processing engine for SPARQL endpoints. In: Aroyo, L., et al. (eds.) ISWC 2011. LNCS, vol. 7031, pp. 18–34. Springer, Heidelberg (2011). https://doi. org/10.1007/978-3-642-25073-6_2
2. Beek, W., Fernández, J.D., Verborgh, R.: LOD-a-lot: a single-file enabler for data science. In: Proceedings of the 13th International Conference on Semantic Systems, SEMANTICS 2017, Amsterdam, The Netherlands, 11–14 Sept 2017, pp. 181–184 (2017)
3. Charalambidis, A., Troumpoukis, A., Konstantopoulos, S.: SemaGrow: optimizing federated sparql queries. In: Proceedings of the 11th International Conference on Semantic Systems, pp. 121–128. ACM (2015)
4. Endris, K.M., Almhithawi, Z., Lytra, I., Vidal, M.-E., Auer, S.: BOUNCER: privacy-aware query processing over federations of rdf datasets. In: Hartmann, S., Ma, H., Hameurlain, A., Pernul, G., Wagner, R.R. (eds.) DEXA 2018. LNCS, vol. 11029, pp. 69–84. Springer, Cham (2018). https://doi.org/10.1007/978-3-319-98809-2_5
5. Endris, K.M., Galkin, M., Lytra, I., Mami, M.N., Vidal, M.-E., Auer, S.: MULDER: querying the linked data web by bridging rdf molecule templates. In: Benslimane, D., Damiani, E., Grosky, W.I., Hameurlain, A., Sheth, A., Wagner, R.R. (eds.) DEXA 2017. LNCS, vol. 10438, pp. 3–18. Springer, Cham (2017). https://doi.org/ 10.1007/978-3-319-64468-4_1
6. Fundulaki, I., Auer, S.: Linked open data - introduction to the special theme. ERCIM News **2014**(96) (2014)
7. Görlitz, O., Staab, S.: SPLENDID: SPARQL endpoint federation exploiting VOID descriptions. In: COLD (2011)
8. Schwarte, A., Haase, P., Hose, K., Schenkel, R., Schmidt, M.: FedX: optimization techniques for federated query processing on linked data. In: Aroyo, L., et al. (eds.) ISWC 2011. LNCS, vol. 7031, pp. 601–616. Springer, Heidelberg (2011). https:// doi.org/10.1007/978-3-642-25073-6_38
9. Vidal, M., Castillo, S., Acosta, M., Montoya, G., Palma, G.: On the selection of SPARQL endpoints to efficiently execute federated SPARQL queries. Trans. Large-Scale Data-Knowl.-Centered Syst. **25**, 109–149 (2016)

Poster Paper
Data Integration for Supporting Biomedical Knowledge Graph Creation at Large-Scale

Samaneh Jozashoori[1,2(✉)] ⓘ, Tatiana Novikova[3] ⓘ, and Maria-Esther Vidal[1,2] ⓘ

[1] L3S Institute, Leibniz University of Hannover, Hannover, Germany
jozashoori@l3s.de
[2] TIB Leibniz Information Centre for Science and Technology, Hannover, Germany
maria.vidal@tib.eu
[3] University of Bonn, Bonn, Germany
s6tanovi@uni-bonn.de

Abstract. In recent years, following FAIR and open data principles, the number of available big data including biomedical data has been increased exponentially. In order to extract knowledge, these data should be curated, integrated, and semantically described. Accordingly, several semantic integration techniques have been developed; albeit effective, they may suffer from scalability in terms of different properties of big data. Even scaled-up approaches may be highly costly due to performing tasks of semantification, curation, and integration independently. To overcome these issues, we devise ConMap, a semantic integration approach which exploits knowledge encoded in ontologies to describe mapping rules in a way that performs all these tasks at the same time. The empirical evaluation of ConMap performed on different data sets shows that ConMap can significantly reduce the time required for knowledge graph creation by up to 70% of the time that is consumed following a traditional approach. Accordingly, the experimental results suggest that ConMap can be a semantic data integration solution that embody FAIR principles specifically in terms of interoperability.

1 Introduction

With the rapid advances in different techniques in the biomedical domain such as Next Generation Sequencing [9], which allow for producing massive amounts of data in acceptable time, and access policies such as FAIR [10] and open data principles, big data has become a quotidian occurrence. However, knowledge discovery from big data, as the criteria to make decisions and take actions, is still a challenging problem. In order to extract knowledge, data residing in different sources should be curated, integrated, and semantically described.

In recent years, the development of Semantic Web technologies with the main purpose of describing the meaning of data in a machine readable fashion,

S. Auer and M.-E. Vidal (Eds.): DILS 2018, LNBI 11371, pp. 91–96, 2019.
https://doi.org/10.1007/978-3-030-06016-9_9

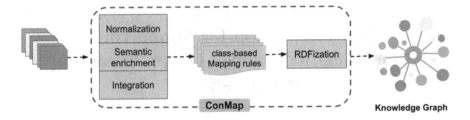

Fig. 1. The ConMap approach. ConMap receives structured data sets from heterogeneous sources as input, and produces a knowledge graph. It relies on conceptual or class-based mapping approach in performing all tasks of semantic enrichment, integration, and transformation i.e. both semantification and integration are performed during class-based mapping and afterwards, based on generated mapping rules, normalized data is transformed as RDF model into the knowledge graph.

has facilitated the implementation of various semantic integration applications. Existing semantic data integration approaches rely on a common framework that allows for the transformation of data in various raw formats into a common data model, e.g., RDF [8]. The mapping rules for data transformation are expressed using mapping languages such as RML [1]. Accordingly, several semantic data integration approaches and tools have been introduced following these technologies such as Karma[1], MINTE [2], SILK [4], and Sieve [7]. Albeit effective, existing semantic data integration tools may suffer from scalability in terms of the dominant dimensions of big data, i.e., volume, variety, veracity, velocity, and value. In fact, even scaled-up approaches are mainly a trade-off between mentioned aspects of big data since it would be highly costly to scale up all the tasks of semantification and integration in terms of more than one dimension, particularly, while being performed independently. More precisely, performing the tasks sequentially results in going through the same procedure of cleaning, semantifying, curating, and transforming for each single data set while their data overlap partially. Moreover, during each mentioned step, the volume of data may be grown and consequently the computational complexity of the next task in the queue, and eventually integration as the last step, will be considerably increased. To overcome drawbacks of existing approaches, we introduce ConMap, a semantic integration approach for big data.

ConMap exploits knowledge encoded in a global schema to perform all the mentioned tasks, i.e., semantification, integration, and transformation in one single step. Therefore, ConMap can be considered as a scalable solution for semantic integration of big data. We have performed an initial experiment study over data sets of various sizes. The observed results suggest that ConMap reduces RDFization time [5] i.e., the time required for transforming heterogeneous structured data sets into RDF.

The rest of the paper is structured as follows: in Sect. 2 the general idea of ConMap is presented as well as detailed explanation of ConMap architecture and

---

[1] http://usc-isi-i2.github.io/karma/.

components. In Sect. 3, the experiments that are performed between ConMap and the attribute-based mapping approach are described and the results are evaluated in terms of time complexity. Finally, Sect. 4 represents our conclusions.

2 The ConMap Approach

ConMap is a semantic data integration approach able to use mapping rules not only for data semantification, but also for curation and integration. ConMap implements a class-based mapping paradigm that resembles the Global-As-View [3] approach of data integration systems [6]; it enables the definition of the mapping, curation, and integration rules per each class in the global schema. Thus, ConMap executes all the tasks, i.e., semantification, curation, and integration at the same time by evaluating these class-based rules. Figure 1 devises the ConMap architecture. ConMap receives real-world data source(s) that represent the same concepts in the global schema but in different formats; it outputs a knowledge graph where input data is integrated and described in a structured way. Data related to each class is extracted from different data sources which are normalized in advance to reduce data redundancies. Afterwards, normalized data is semantified in order to describe and integrate this data in the knowledge graph. The components of ConMap can be summarized as below:

- **Normalization:** To overcome interoperability issues, all data sets are normalized considering the real world concepts. Since each concept may be represented by more than one data source and each data source may involve more than one concept, the process of normalization is based on the decomposition of each data set in terms of the global schema classes.
- **Semantic Enrichment:** Since the mapping process enables semantification and curation of data, the approach that is applied for mapping plays a significant role both on computational complexity and the quality of semantified data. The attribute-based mapping approach is source-oriented which means it semantifies the concepts that are presented in each source based on the attributes that are available in the same data source. In contrast, the class-based mapping approach is concept-oriented; it defines semantic descriptions of each concept according to the attributes that are determined by a global schema and expressed by variant data sources either equally or differentially.
- **Integration:** To integrate data residing at heterogeneous sources, they all required to be transformed into a unified representation. Since in ConMap approach, in order to decrease the cost of data comparison, data integration precedes data transformation, semantic descriptions provided during the generation of mapping rules are employed to translate different representations of data into a unified one. Furthermore, semantic descriptions provide actionable information for data fusion in case of inconsistency of data values between different sources.
- **RDFization:** The last component to be executed in ConMap is RDFization. It is important to note that despite being the last step, the performance of previous steps can be also observed through the outcome of this component

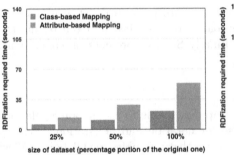
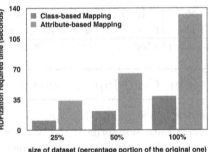

(a) RDFizing one class including 5 attributes from three different sized data sets

(b) RDFizing one class including 12 attributes from three different sized data sets

Fig. 2. Experimental results. (a) The required time for RDFization of one class including five attributes from three different sized data sets. (b) The required time for RDFization of one class including twelve attributes from three different sized data sets.

which evaluates class-based mapping rules for transforming normalized and semantified data sets into the knowledge graph.

3   Experimental Study

In this paper, the performance of two mapping paradigms are compared: the class-based mapping approach provided by ConMap, and an attribute-based approach which is commonly followed by existing tools, e.g., Karma. We address two research questions: **(RQ1)** Does ConMap reduce the time complexity of RDFization? **(RQ2)** How influential a mapping approach can be in terms of execution time when the complexity of the class increases?

Benchmark: In this study, a data set with overall size of 169.8 MB is extracted from COSMIC[2], an online database of somatic mutations that are found in human cancer. The data set is in tab separated format comprising 557,162 records of lung cancer related coding point mutations that are derived from targeted and genome wide screens.

Metrics: The behavior of the studied mapping approaches is evaluated by measuring the execution time in seconds for transforming a data set into RDF applying that approach.

Implementation: The mapping rules[3] are expressed in the RML mapping language. The RDFization is implemented in Python 3.6. The experiment was executed on a Ubuntu 17.10 (64 bits) machine with Intel W-2133, CPU 3.6 GHz, 1 physical processor; 6 cores; 12 threads, and 64 GB RAM.

---

[2] https://cancer.sanger.ac.uk/cosmic.
[3] https://github.com/samiscoding/DILS.

Experimental Setup: Two experiments are set up in this study: (**E1**) In order to better understand the influence of mapping approach on time complexity of RDFization, the experiment is run on three different sized data sets: the first one is the preprocessed data set derived from the original mutation data set that is extracted from COSMIC without any decrease regarding its size while the two other data sets are extracted from the first one. The records that are included in two latest data sets are 50% and 25% randomly selected records of the first data set. The result of this experiment is shown in Fig. 2(a). (**E2**) To study how time complexity of each mapping approach fluctuated with the increase in the number of attributes for a class, for each mapping approach two separated sets of mapping rules are defined; one mapping rule set for an RDF class with twelve attributes and the other one including five attributes. The experimental results can be seen in Fig. 2(b). Based on the results of explained experiments that are illustrated in Fig. 2, the execution time increases in case of using the attribute-based mapping rules for transformation of data in both sets including different numbers of attributes which positively answers the **RQ1**. Moreover, the observed results lead to answer **RQ2** as follows: in attribute-based mapping approach, the required execution time for transforming one class of data will grow when the number of its attributes increases, however, in class-based mapping the time complexity is not a function of class complexity.

3.1   Discussion

The evaluation results can be simply explained according to the fact that the attribute-based mapping approach performs the same procedure of creating *subject-predicate-object* triple for every single attribute of a class. In contrast, the class-based mapping approach transforms each concept or class including all its attributes to one RDF class in a single run. Therefore, class-based mapping approach can be considered as a fundamental procedure for transforming raw data into RDF model in an integrated non-redundant way.

4   Conclusions and Future Work

We introduced ConMap, a semantic integration approach that deploys the knowledge encoded in an ontology to perform semantification and integration during transformation, in a way that a big data scalability can be acquired. We empirically showed that ConMap can reduce the execution time of semantic integration of structured data sets into a knowledge graph. Although experimental results demonstrated in this paper were derived by all components of ConMap, there is still room to illustrate the power of this approach in terms of integration. There exist open problems regarding different dimensions of big data that can be tackled during the integration process in ConMap; they include veracity which refers to noise, abnormality and inconsistency of available data. In our future work we will more focus on improving ConMap from data fusion perspective.

Acknowledgement. This work has been supported by the European Union's Horizon 2020 Research and Innovation Program for the project iASiS with grant agreement No 727658.

References

1. Dimou, A., Vander Sande, M., Colpaert, P., Verborgh, R., Mannens, E., Van de Walle, R.: RML: a generic language for integrated RDF mappings of heterogeneous data. In: LDOW (2014)
2. Endris, K.M., Galkin, M., Lytra, I., Mami, M.N., Vidal, M.-E., Auer, S.: MULDER: querying the linked data web by bridging RDF molecule templates. In: Benslimane, D., Damiani, E., Grosky, W., Hameurlain, A., Sheth, A., Wagner, R.R. (eds.) DEXA 2017. LNCS, vol. 10438, pp. 3–18. Springer, Cham (2017). https://doi.org/10.1007/978-3-319-64468-4_1
3. Friedman, M., Levy, A.Y., Millstein, T.D., et al.: Navigational plans for data integration. AAAI/IAAI **1999**, 67–73 (1999)
4. Isele, R., Bizer, C.: Active learning of expressive linkage rules using genetic programming. Web Semant.: Sci., Serv. Agents World Wide Web **23**, 2–15 (2013)
5. Jha, A., et al.: Towards precision medicine: discovering novel gynecological cancer biomarkers and pathways using linked data. J. Biomed. Semant. **8**(1), 40 (2017)
6. Lenzerini, M.: Data integration: a theoretical perspective. In: Proceedings of the Twenty-First ACM SIGMOD-SIGACT-SIGART Symposium on Principles of Database Systems, pp. 233–246. ACM (2002)
7. Mendes, P.N., Mühleisen, H., Bizer, C.: Sieve: linked data quality assessment and fusion. In: Proceedings of the 2012 Joint EDBT/ICDT Workshops, pp. 116–123. ACM (2012)
8. Miller, E.: An introduction to the resource description framework. Bull. Am. Soc. Inf. Sci. Technol. **25**(1), 15–19 (1998)
9. Reis-Filho, J.S.: Next-generation sequencing. Breast Cancer Res. **11**(3), S12 (2009)
10. Wilkinson, M.D., et al.: The fair guiding principles for scientific data management and stewardship. Sci. Data **3** (2016)

DISBi: A Flexible Framework for Integrating Systems Biology Data

Rüdiger Busche[1]($\boxtimes$)(iD), Henning Dannheim[2], and Dietmar Schomburg[2](iD)

[1] Institute of Cognitive Science, Osnabrück University, Osnabrück, Germany
rbusche@uos.de
[2] Department for Bioinformatics and Biochemistry,
Technische Universität Braunschweig, Braunschweig, Germany
{h.dannheimm,d.schomburg}@tu-braunschweig.de

Abstract. Systems biology aims at understanding an organism in its entirety. This objective can only be achieved with the joint effort of specialized work groups. These collaborating groups need a centralized platform for data exchange. Instead data is often uncoordinatedly managed using heterogeneous data formats. Such circumstances present a major hindrance to gaining a global understanding of the data and to automating analysis routines.

DISBi is a framework for creating an integrated online environment that solves these problems. It enables researchers to filter, integrate and analyze data directly in the browser. A DISBi application dynamically adapts to its data model. Thus DISBi offers a solution for a wide range of systems biology projects.

An example installation is available at disbi.org. Source code and documentation are available from https://github.com/DISBi/django-disbi.

Keywords: Systems biology · Data integration · Data exchange

1 Introduction

In systems biology, researchers try to gain a system-level understanding of an entire organism. This objective requires investigating organisms from different perspectives involving diverse experimental and computational approaches [6]. As the different approaches usually require specialization, work groups bundle their efforts in consortia, each work group investigating a different level of biological information. The huge amounts of data generated in different experimental and simulation approaches need to be integrated to reveal the underlying patterns. While data exchange is a crucial aspect of this effort, it is often hindered by a lack of standardized data formats and a centralized data storage. Though different standards exist [10], they are often not used consistently throughout the project due to inflexibility of the formats or diverging preferences of different work groups [7]. These circumstances greatly hinder the progress towards

© Springer Nature Switzerland AG 2019
S. Auer and M.-E. Vidal (Eds.): DILS 2018, LNBI 11371, pp. 97–102, 2019.
https://doi.org/10.1007/978-3-030-06016-9_10

system-level understanding, as complete data are seldomly available and often need to be integrated manually.

With DISBi, we present a framework to easily construct an online environment for *Data Integration* in *Systems Biology*. It enables users to manage, organize and integrate heterogeneous data generated in systems biology projects. Existing systems like *Aureolib* [3] rely on fixed data models, which makes the applications useful only for a narrow range of projects. Often these applications are limited to manage only data from a single organism or a single experimental method. Instead, DISBi's approach is to generalize the problem of data integration in systems biology. Thus DISBi is a framework rather than an application that allows users to set up a web app customized for the data of a particular project. Programmers can focus on the logic relevant for the project, rather than implementing details. Researchers profit from the availability of integrated data throughout the course of a project.

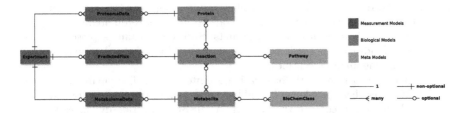

Fig. 1. Exemplary layout of the abstract data model. A possible data model for integrating proteome, flux balance analysis and metabolome data. Reactions are grouped in pathways and metabolites in biochemical classes by utilizing meta models. Every measurement model has a non-optional relation to one *Experiment* and one biological model. Note that this is only one possible data model for a DISBi application. The abstract data model can accommodate varying numbers biological and measurement models with more complex relationships.

2   Data Model

When setting up a new instance of a DISBi application for a systems biology project, only a new data model has to be defined. This data model should capture all information relevant to the project. The defined data model needs to adhere to DISBi's *abstract data model*, which is a relational data model with additional constraints. These constraints serve the purpose of imposing some structure on the data model that can be exploited by the framework, while giving the developer the greatest possible freedom in modeling the domain of the project. In the abstract data model, data is split into three categories: *biological models*, *measurement models*, *meta models* (Fig. 1). While these are abstract categories a concrete *Experiment* model is necessarily part of the data model as well.

Every data source (simulation or wet-lab experiment) is stored in the Experiment model. The Experiment model should include all meaningful experimental parameters, but can also accommodate free input labels. What exactly these parameters need to be ultimately depends on the design of the investigation. We find it useful to include only experimental parameters that were varied across the investigation, e.g. carbon source, and meta data such as the experimental method used to generate the data.

Measurement models store data points generated in an experiments, such as the response from a mass spectrometer. Biological models store the biological entities the data points map to, such as proteins or metabolites. Meta models store information about these biological entities, such as pathways or functional groups. The data model needs to be designed such that each instance of a measurement model can uniquely be identified by its relation to an experiment and a biological object. Thus it is representing data about a certain biological object measured in a certain experiment. The biological models can be related in arbitrary ways, but must not create cyclic relations. For example a protein maps to a reaction that maps to a metabolite. These relations define an implicit relation between proteins and metabolites, but an additional explicit relation between a reaction and a metabolite would introduce the possibility of inconsistency and is therefore forbidden by the abstract data model.

3   Data Integration

Data of biological models and their relations have to be determined and uploaded to the system before it is used for integrating experimental data. These data form the backbone of the data integration process and can be conveniently uploaded through the admin interface in tabular formats. Relations of the biological models are included in the tabular data by including columns of unique identifiers of related biological objects. The naming and relations of the biological objects should be agreed upon by all partners in the project as it is only possible to upload measurements that map to biological objects that are included in the system. Thus the biological objects serve as a kind of controlled vocabulary. Adapting the data model during the project can be achieved by using Django's built-in migration framework.

Data integration is done based on the relations of the biological models by combining the results from all matched experiments in a dynamic table. The relations between the models are dynamically inspected at runtime. Data points from different experiments that map to the same biological object as well as data points that map to related biological entities are combined in one row. For example, transcriptome and proteome data will be presented together, if the respective protein derives from the respective gene. This format makes it easy to analyze the data for correlations between related biological entities and to test predictions from simulations. Meta models play no role in data integration, but simply function as a container for data that cannot be put on the biological models due to normalization constraints. That means that while the information from meta models is included in the dynamic table, biological objects are not joined together based on information in the meta models.

4   Usage

A DISBi application is split into two main interfaces. The *Filter View* provides the user with a tool to find experiments of interest based on experimental parameters. It is automatically constructed from the underlying data model. The user is shown a preview of the matched experiments, which allows for interactive exploration of available experiments. Once a set of experiments of interest is determined, the user is taken to the *Data View*. In this view, the integrated experimental data is presented in a table and can be further filtered, exported and analysed. The analysis capabilities include calculating fold changes, plotting distributions of single columns and comparing two experiments in a scatter plot. These functionalities provide the users with a tool to quickly get an overview of the datasets and determine which data are worthwhile for in-depth analysis.

The DISBi framework supports an easy to use, customizable admin interface that facilitates uploads of large datasets from common file formats such as Excel, CSV or JSON.

Extended version of Django model classes are used to define the data model. This provides a high-level interface for specifying the data base scheme that does not require a deep understanding of the underlying database structures and can be accomplished by everyone with basic Python proficiency. As the abstract data model puts no constraint on the number of models or fields, it can accommodate a great variety of different project outlines.

5   Experiences

DISBi was successfully applied to internal projects in the Department for Bioinformatics and Biochemistry at the Technische Universität Braunschweig for integrating data from the organisms *Chlostridium difficile*, *Aromatoleum aromaticum* EbN1 and *Dinoroseobacter shibae* as well as for public data from *Sulfolobus solfataricus* [12]. The integrated data sources include transcriptome, proteome, metabolome and predicted metabolic flux data. The integrated methods range from RNA sequencing and mass spectrometry to microarray data and simulations. This demonstrates the applicability of DISBi to a wide range of different data.

6   Implementation

The DISBi framework is designed with Python 3.6 (python.org) as backend language based on the Django web development framework (v1.11 djangoproject.com). PostgreSQL (postgresql.org) is used as database backend. Interactive data visualization is implemented with NumPy [11] and matplotlib [4].

The jQuery (jquery.com) JavaScript library and SASS (sass-lang.com) pre-processor are used to provide a responsive user interface with a consistent visual appearance.

By using only open source technologies, we ensure that the DISBi framework is freely available and extendable. Moreover, by using Python future developers of DISBi will have access to the vast ecosystem of scientific software available in Python [5] for extending DISBi with new features.

7   Related Work

Many systems exist that tackle integrating heterogenous data in multi work group system biology projects [14]. The most prominent systems are DERIVA [2], openBIS [1] and the SEEK platform [13] together with the ISA toolchain [8]. These platforms are well established and go far beyond the current scope of DISBi. They offer functionality for automatically uploading data to the system when they are produced at the measurement device and ultimately enable the user to deploy integrated data in public repositories. In contrast to DISBi they focus on *asset management*, i.e. attaching meta data to arbitrary data sources and storing these data sources in a unified fashion. DISBi is more focused on establishing correspondences between single data points. Hence, larger systems such as DERIVA can accommodate more heterogeneous data and are therefore applicable to very large projects, while DISBi is restricted to tabular data.

DISBi should therefore be seen as a lightweight complementary approach that can be used to integrate managed assets in a more fine-grained manner.

The importance of having the system dynamically adapt to its data model is highlighted in the design of DERIVA [9]. This ensures that the system is applicable to a wide variety of different projects. DISBi follows the same design philosophy.

8   Conclusion

With DISBi, we present a powerful framework for the construction of custom data integration platforms for systems biology projects. Its flexibility and customizability render it an applicable solution for managing data in small to mid-sized projects and making data publicly available. The source code is freely available under terms of the MIT license, which allows other developers to modify and evolve the software. We will continue to explore possible ways of automating data integration and analysis based on the abstract data model.

Acknowledgements. The authors thank Meina Neumann-Schaal for critical reading of the manuscript and four anonymous reviewer for their instructive comments. Rüdiger Busche thanks Pascal Nieters for support in the publication process.

This work was supported by the Federal State of Lower Saxony, Niedersächsisches Vorab (VWZN2889)/3215.

References

1. Bauch, A., et al.: openBIS: a flexible framework for managing and analyzing complex data in biology research. BMC Bioinform. **12**(1), 468 (2011). https://doi.org/10.1186/1471-2105-12-468

2. Bugacov, A., Czajkowski, K., Kesselman, C., Kumar, A., Schuler, R.E., Tangmunarunkit, H.: Experiences with DERIVA: an asset management platform for accelerating eScience. In: Proceedings of the 13th IEEE International Conference on eScience, eScience 2017, pp. 79–88 (2017). https://doi.org/10.1109/eScience.2017.20

3. Fuchs, S., et al.: Aureolib - a proteome signature library: towards an understanding of staphylococcus aureus pathophysiology. PLoS One **8**(8), e70669 (2013). https://doi.org/10.1371/journal.pone.0070669

4. Hunter, J.D.: Matplotlib: a 2D graphics environment. Comput. Sci. Eng. **9**(3), 99–104 (2007). https://doi.org/10.1109/MCSE.2007.55

5. Jones, E., Oliphant, T., Peterson, P.: SciPy: open source scientific tools for Python (2014)

6. Kitano, H.: Systems biology: a brief overview. Science **295**(5560), 1662–1664 (2002). https://doi.org/10.1126/science.1069492

7. Lubitz, T., Hahn, J., Bergmann, F.T., Noor, E., Klipp, E., Liebermeister, W.: SBtab: a flexible table format for data exchange in systems biology. Bioinformatics **32**(16), 2559–2561 (2016). https://doi.org/10.1093/bioinformatics/btw179

8. Rocca-Serra, P., et al.: ISA software suite: supporting standards-compliant experimental annotation and enabling curation at the community level. Bioinformatics **27**, 2354–2356 (2011). https://doi.org/10.1093/bioinformatics/btq415

9. Schuler, R.E., Kesselman, C., Czajkowski, K.: Accelerating data-driven discovery with scientific asset management. In: 2016 IEEE 12th International Conference on e-Science (e-Science), pp. 31–40. IEEE (2016). https://doi.org/10.1109/eScience.2016.7870883

10. Taylor, C.F., et al.: Promoting coherent minimum reporting guidelines for biological and biomedical investigations: the MIBBI project. Nat. Biotechnol. **26**(8), 889–896 (2008). https://doi.org/10.1038/nbt.1411

11. Van Der Walt, S., Colbert, S.C., Varoquaux, G.: The NumPy array: a structure for efficient numerical computation. Comput. Sci. Eng. **13**(2), 22–30 (2011). https://doi.org/10.1109/MCSE.2011.37

12. Wolf, J., et al.: A systems biology approach reveals major metabolic changes in the thermoacidophilic archaeon Sulfolobus solfataricus in response to the carbon source L-fucose versus D-glucose. Mol. Microbiol. **102**(5), 882–908 (2016). https://doi.org/10.1111/mmi.13498

13. Wolstencroft, K., et al.: The SEEK: a platform for sharing data and models in systems biology. Methods Enzymol. **500**, 629–655 (2011). https://doi.org/10.1016/B978-0-12-385118-5.00029-3

14. Wruck, W., Peuker, M., Regenbrecht, C.R.A.: Data management strategies for multinational large-scale systems biology projects. Brief. Bioinform. **15**(1), 65–78 (2014). https://doi.org/10.1093/bib/bbs064

Biomedical Data Analytics

Biomedical Data Analysis

Using Machine Learning to Distinguish Infected from Non-infected Subjects at an Early Stage Based on Viral Inoculation

Ghanshyam Verma[1,2(✉)] (iD), Alokkumar Jha[1,2],
Dietrich Rebholz-Schuhmann[1,2,3], and Michael G. Madden[1,2] (iD)

[1] Insight Centre for Data Analytics, National University of Ireland Galway,
Galway, Ireland
{ghanshyam.verma,alokkumar.jha,rebholz}@insight-centre.org
[2] School of Computer Science, National University of Ireland Galway,
Galway, Ireland
michael.madden@nuigalway.ie
[3] ZB MED - Information Center for Life Sciences, University of Cologne,
Cologne, Germany

Abstract. Gene expression profiles help to capture the functional state in the body and to determine dysfunctional conditions in individuals. In principle, respiratory and other viral infections can be judged from blood samples; however, it has not yet been determined which genetic expression levels are predictive, in particular for the early transition states of the disease onset. For these reasons, we analyse the expression levels of infected and non-infected individuals to determine genes (potential biomarkers) which are active during the progression of the disease. We use machine learning (ML) classification algorithms to determine the state of respiratory viral infections in humans exploiting time-dependent gene expression measurements; the study comprises four respiratory viruses (H1N1, H3N2, RSV, and HRV), seven distinct clinical studies and 104 healthy test candidates involved overall. From the overall set of 12,023 genes, we identified the 10 top-ranked genes which proved to be most discriminatory with regards to prediction of the infection state. Our two models focus on the time stamp nearest to $t = 48$ hours and nearest to $t =$ "*Onset Time*" denoting the symptom onset (at different time points) according to the candidate's specific immune system response to the viral infection. We evaluated algorithms including k-Nearest Neighbour (k-NN), Random Forest, linear Support Vector Machine (SVM), and SVM with radial basis function (RBF) kernel, in order to classify whether the gene expression sample collected at early time point t is infected or not infected. The "*Onset Time*" appears to play a vital role in prediction and identification of ten most discriminatory genes.

Keywords: Machine learning · Respiratory viral infection
Prediction · Deferentially expressed genes

© Springer Nature Switzerland AG 2019
S. Auer and M.-E. Vidal (Eds.): DILS 2018, LNBI 11371, pp. 105–121, 2019.
https://doi.org/10.1007/978-3-030-06016-9_11

1   Introduction

Respiratory viral infections are common diseases which are caused by a wide range of viruses, e.g., H1N1, H3N2, RSV and HRV, affecting the respiratory tract. While patients usually recover in a short period of time without any treatment, respiratory viral infections can lead to severe outcomes among individuals with other aggravating primary diseases, in particular, when these are deleterious to the function of the respiratory system. Such severe cases may increase the likelihood of death in elderly or immuno-compromised individuals [14]. Moreover, each influenza epidemic leads to an increase in healthcare costs through excessive hospitalizations apart from the need for substantial amounts of vaccines, and the spread of respiratory virus diseases affect all age groups and thus can lead to periodic epidemics [25]. Overall, the early identification of respiratory viral infections could be useful as a means to reduce large-scale outbreaks and periodic epidemics as well as achieving early intervention for individual patients [13].

In this paper, we investigated the changes in gene expression distinguishing infected individuals from non-infected ones. We use different ML methods to determine the most predictive changes comparing samples from healthy and infected individuals, using public data collected in seven different studies involving healthy individuals before and after inoculation of the viruses. This data (gene expression only) – generated from these seven challenge studies – has been released in 2016 and is available on Gene Expression Omnibus (GEO). In 2017, the label information (non-infected vs. infected) associated with this dataset also had been made available for open access to all. We use this label information as a ground-truth for labeling the whole data. ML solutions form a vital role in the identification of specific patterns, and subsequent functional annotation of the identified genes can explain the causality behind the exposed patterns. Gene expression changes often happen due to some regulatory markers, while other genes behave as housekeeping genes. Therefore, identification of relevant patterns and responsible regulatory markers at consistent time points should yields credible biomarkers in such cases. In this work we identify top ten such biomarkers which are found to be highly contributing in progression of respiratory viral infections at an early stage. The labeled data with code and build ML models are available here: https://github.com/GhanshyamVerma/DILS_2018.

2   The Respiratory Viral Data Sets

We conducted ML experiments on the data collected from 7 Respiratory Viral Challenge studies which is available for open access on GEO (accession number GSE73072)[1]. These respiratory viral challenge studies consist of a total of 151 human volunteers, each of whom was exposed to one of 4 viruses, summarised in Table 1 [12].

In Table 1, the first column represents the sub-study designation, the second column denotes the type of virus used in the challenge, the third and the

---

[1] https://www.ncbi.nlm.nih.gov/geo/query/acc.cgi?acc=GSE73072.

Table 1. Details of the data collected in the seven respiratory virus challenge studies [12]

Challenge	Virus	Year	Location	IRB Protocol	Duration (hrs)	#Subjects	#Time-points
DEE1	RSV	2008	Retroscreen	Pro00002796	166	20	21
DEE2	H3N2	2009	Retroscreen	Pro00006750	166	17	21
DEE3	H1N1	2009	Retroscreen	Pro00018132	166	24	20
DEE4	H1N1	2010	Retroscreen	Pro00019238	166	19	21
DEE5	H3N2	2011	Retroscreen	Pro00029521	680	21	23
HRV UVA	HRV	2008	UoVirginia	Pro00003477	120	20	15
HRV Duke	HRV	2010	Duke Univ.	Pro00022448	136	30	19

fourth columns represent the year and the location of the conducted sub-study, respectively, the fifth column represents the DUHS IRB protocol number, the sixth column represents the duration of the sub-study in hours and the last two columns denote the number of subjects and the number of time-points collected per subject, respectively [12].

All the participants were healthy when they enrolled for the study. After enrolment in the study, all subjects were inoculated with one of the 4 viruses (H1N1, H3N2, HRV, RSV). Their blood samples were taken at different pre-defined time-points, thus delivering samples from non-infected individuals as well as from infected ones. The samples from non-infected individuals were taken at two time-points before the inoculation of the virus, as shown in Fig. 1 (inspired by a figure by Liu et al. [12]). All the subjects were exposed to the virus immediately

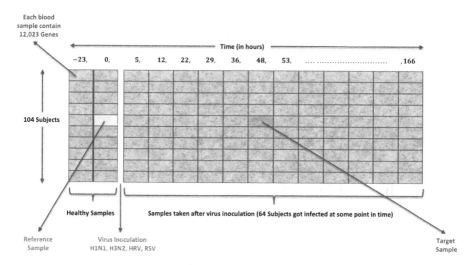

Fig. 1. Layout describing characteristics of the data. Every cell depicting blood sample taken at some point of time during the whole study and contains gene expression values of 12,023 human genes.

after taking the healthy blood sample (at time-point 0). During each study, blood samples were taken for twice before the inoculation of virus and at various time stamps after the inoculation of virus. The whole blood gene expression data was obtained using Affymetrix Human U133A 2.0 GeneChips. Additional details can be found on GEO (accession number GSE73072).

From the start, 151 subjects were enrolled in the 7 challenge studies, however, we had to exclude 47 subjects from the study. Among those 47 subjects, 44 subjects had inconsistencies between their declared symptomatic status and the measured shedding status (see Table 2). These 44 clinically ambiguous subjects were at some time either acutely infected but remained asymptomatic or were not infected but did turn acutely symptomatic [12], therefore, we must conclude that these 44 subjects data is inconsistent (faulty). We cannot draw any conclusions from faulty data. Moreover, faulty data can be misleading and harmful while model building. Apart from these 44 subjects, three more subjects have been excluded because there is no Affymetrix data available for them (subjects 6, 9 and 21 from the HRV Duke university sub study). We have identified those 47 ambiguous subjects whose data is faulty, removed them and the unambiguous labeled data with code and build ML models can be accessed using a link provided in the Introduction section.

Table 2. Detail of the ambiguous subjects those excluded due to inconsistencies between their declared symptomatic status and measured shedding status.

Sr. No.	Challenge	Subject IDs (Ambiguous subjects)	Total (Ambiguous subjects)
1	DEE1	13, 15, 16	3
2	DEE2	2, 4	2
3	DEE3	1, 2, 5, 7, 11, 15, 18, 21, 23	9
4	DEE4	5, 7, 8, 9, 10, 11, 12, 13, 17, 19	10
5	DEE5	3, 7, 15, 16, 17	5
6	UoVirginia	1, 10, 12, 17	4
7	Duke Univ.	3, 10, 11, 15, 18, 20, 25, 27, 28, 29, 30	11
			44

3   Experimental Design

The overall goal of our study is to analyze the ability of different ML algorithms to predict the state of health and disease soon after the disease onset time. As we do not have always data for each subject exactly the times we are interested in, we have taken nearest available time-point. We make the assumption that at $t = 0$, just before the inoculation, all the candidates are not sick and at $t = 48\,h$ (approximately) or at the onset time, the effect of virus inoculation should be visible, and thus exposed in the gene expression data. After first mining for expression patterns, we are also interested in finding the important genes/biomarkers which are highly likely to contribute to the progression of the respiratory viral infection.

After excluding the 47 ambiguous candidates, we were left with total 104 candidates, all of whom were healthy at $t = 0$. Out of these 104 healthy subjects, 64 became sick at some point of time after inoculation of the virus and the other 40 remained healthy during the whole study. There is no onset time for these 40 non-infected subjects, therefore, we took the average for the available onset values, which was 55.01 h after inoculation.

Our main focus is on the gene expression levels when comparing the 40 subjects who did not become infected after inoculation with the 64 who did. We designed 4 different experiments: for each experiment we made different use of the number of subjects that got infected after inoculation and of the time-points (48 h vs. onset time).

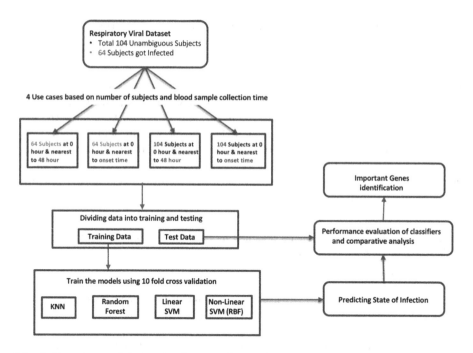

Fig. 2. Experimental design for distinguishing infected subjects form non-infected subjects by exploiting ML algorithm's ability to learn pattern.

We believe that our experiments play a useful role in determining the involvement of particular genes in the states of infection at the early stage of the disease. We took the data of all the unambiguous subjects and divided it into four subsets as shown in Fig. 2. The details of the four experiments designed using these

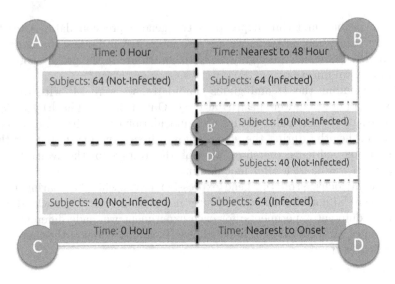

Fig. 3. A view of 4 experiments that comprises significant part of the overall experimental design.

four subsets of data can be seen in Figs. 3 and 4. For each of the four subsets of the data, we partitioned the data into training and test sets, and then applied four well established ML approaches. The random sampling is done with preserving the class distribution to partition the whole data into training and test. In order to reduce the risk of overfitting we have applied 10-fold cross validation, repeated 3 times. The build model were then used to predict the state of infection for the kept test data. Finally, the performance evaluation and important gene identification steps have been carried out.

Table 3. Detail of the experiments designed by combining two or more states of subject's status of gene expression profile.

Experiment no.	States	Description
1	$(A + B)$	64 subjects data collected at 0 and nearest to 48 h
2	$(A + D)$	64 subjects data collected at 0 and nearest to onset time
3	$(A + B + B' + C)$	104 subjects data collected at 0 and nearest to 48 h
4	$(A + C + D + D')$	104 subjects data collected at 0 and nearest to onset or average onset time.

We identified 6 sets of data denoting different states, and labelled them State A, B, B', C, D, D' (see Fig. 3). State A contains the gene expression profile of all the 64 subjects which are healthy at time-point 0: these 64 subjects showed clear signs of infection at some point of time after the inoculation of the virus. States

B and D determine the gene expression profiles of the same 64 subjects at the time-point nearest to 48 h or nearest to onset time, respectively. State C shows the gene expression profile of 40 subjects at 0 timestamp: these 40 subjects never get infected throughout the duration of the study. States B' and D' show the gene expression profiles of the same 40 subjects at nearest to 48 h or at nearest to average onset time, respectively.

We carried out four experiments by combining two or more of the above states based on the number of infected and non-infected subjects and timestamps at which their blood samples are collected. These experiments are designed in such a way so that we can analyse the differences in disease prediction at two different early stages and find the most important Differentially Expressed Genes (DEG) across the different timestamps. The details of these experiments are shown in Table 3. The numbers of positive and negative samples for each designed experiment at different time point are shown in Fig. 4.

Time $\rightarrow$	0 Hours		48 Hours		Onset Time	
Experiment $\downarrow$	P	N	P	N	P	N
$(A + B)$	0	64	64	0		
$(A + D)$	0	64			64	0
$(A + B + B' + C)$	0	104	64	40		
$(A + C + D + D')$	0	104			64	40

Fig. 4. Positive and negative sample counts for each experiment at different time points. Here P denotes positive samples (infected) and N denotes negative samples (non-infected).

4 Methodology

In this section we briefly explain the methodology used for classifying the state of health of any individual at any given time point t. It is well-known that no single ML algorithm is best for all kind of datasets, so we tested a selection of different ML approaches. In all experiments, 78% of the data is used for training the classifiers and the remaining 22% is kept as a hold-out test set. The stratified sampling is used to partition the whole data into training and test. To build the ML model for each algorithm we estimated model parameters over the training data using 10-fold cross validation, repeated 3 times.

First, we used the very simple baseline algorithm, k-NN which does not have any in-build capability to deal with high dimensional data [4], however, it can be used to set a base to compare the results and to see the improvements yielded by more complex algorithms. We also used the Random Forest algorithm which is an ensemble technique and has proven to be an efficient approach for the classification of microarray data as well as for gene selection [5]. We then employed both linear SVM [2] and SVM with RBF kernel which has inbuilt capability to learn pattern from high dimensional data [17]. We have used R programming language version 3.4.1 for coding [15].

4.1 k-Nearest Neighbour (k-NN)

k-NN has two stages, the first stage is the determination of the nearest neighbours i.e. the value of k and the second is the prediction of the class label using those neighbours. The "k" nearest neighbours are selected using a distance metric [4]. We have used Euclidean distance for our experiments. This distance metric is then used to determine the number of neighbours. There are various ways to use this distance metric to determine the class of the test sample. The most straightforward way is to assign the class that majority of k-nearest neighbours has. In the present work, the optimum value of k is searched over the range of $k = 1$ to 50. The best value of the parameter k obtained for each experiment can be found in Sect. 5.

4.2   Random Forest

Random Forest is often well-suited for microarray data. It can cope with noisy data and can be used when the number of samples is much smaller than the number of features. Furthermore, it can determine the relevance of variables in the decision process, which can be used for selecting the most relevant genes [5]. It is based on the ensemble of many classification trees [11,18]. Each classification tree is created by selecting a bootstrap sample from the whole training data and a random subset of variables with size denoted as $mtry$ are selected at each split. We have used the recommended value of $mtry : (mtry = \sqrt{(number\ of\ genes)})$ [5]. The number of trees in the ensemble is denoted as $ntree$. We have used $(ntree) = 10,001$ so that each variable can reach a sufficiently large likelihood to participate in forest building as well as in variable importance computations.

4.3   Support Vector Machine (SVM)

Assume that we have given a training set of instance-label pairs $(\boldsymbol{x}_i, y_i); \forall i \in \{1, 2, \ldots, l\}$ where $\boldsymbol{x}_i \in \mathbb{R}^n$ and $\boldsymbol{y} \in \{1, -1\}^l$, then the SVM [2,7,8] can be formulated and solved by the following optimization problem:

$$\min_{\boldsymbol{w}, b, \xi_i} \quad \tfrac{1}{2}\boldsymbol{w}^T\boldsymbol{w} + C \sum_{i=1}^{l} \xi_i,$$

$$\text{subject to} \quad y_i \left(\boldsymbol{w}^T \phi\left(\boldsymbol{x}_i\right) + b\right) \geq 1 - \xi_i,$$

$$\xi_i \geq 0.$$

Here the parameter $C > 0$ is the penalty parameter of the error term [8] and $\xi_i \forall i \in \{1, 2, \ldots, l\}$ are positive slack variables [2]. For linear SVM, we did a search for best value of parameter C for a range of values $(C = 2^{-5}, 2^{-3}, \ldots, 2^{15})$ and the one with the best 10-fold cross validation accuracy has finally been chosen.

We also used SVM with RBF kernel which is a non-linear kernel. There are four basic kernels that are frequently used: linear, polynomial, sigmoid, and RBF. We picked the RBF kernel, as recommended by Hsu et al. [8]. It has the following form:

$$K\left(\boldsymbol{x}_i, \boldsymbol{x}_j\right) = \exp\left(\frac{-\|\boldsymbol{x}_i - \boldsymbol{x}_j\|^2}{2\sigma^2}\right); \frac{1}{2\sigma^2} > 0.$$

We performed a grid-search over the values of C and σ using 10-fold cross validation. The different pairs of (C, σ) values are tried in the range of $(C = 2^{-5}, 2^{-3}, \ldots, 2^{15}; \sigma = 2^{-25}, 2^{-13}, \ldots, 2^3)$ and the values with the best 10-fold cross validation accuracy are picked for the final model building (see Sect. 5).

5   Results

We experimentally obtained the 10-fold cross validation accuracy and hold-out test set accuracy using four algorithms including k-NN, Random Forest, linear SVM, and SVM with RBF Kernel. Based on the results obtained on the hold-out test set for all the four experiments, it can be concluded that the Random Forest model performs better than the rest of the algorithms (see Tables 4, 5, 6 and 7). Random Forest gives the most stable and consistently highest accuracy on the hold-out test set. Moreover, the Random Forest has the additional capability to assign a relevance score to the variables (genes), hence, we have selected random forest for the determination of the genes playing the most important role in the development of the infection.

Table 4. Results on 64 infected subjects data at 0 and nearest to 48 h (Experiment 1).

Sr. No.	Algorithm	Model parameters	Accuracy (10-fold CV)	Accuracy (hold-out)
1	k-NN	$k = 23$	67.66%	53.57%
2	Random forest	$mtry = 109$, $ntree = 10001$	75.33%	64.29%
3	Linear SVM	$C = 0.03125$	68.33%	64.29%
4	SVM with RBF kernel	$C = 5$, $\sigma = 3.051758 \times 10^{-5}$	73%	64.29%

When the 10-fold cross-validation accuracy is considered, none of the algorithms uniformly outperform the others. The SVM with RBF Kernel is able to achieve highest 10-fold cross-validation accuracy for the last 3 experiments, however, the random forest also has similar performance for these last 3 experiments and is even better for the first experiment.

Overall, the results are best when "*Onset Time*" is considered for all 104 subjects (experiment 4) in comparison to the rest of the experiments. This is

Table 5. Results on 64 infected subjects data at 0 and nearest to the onset time (Experiment 2).

Sr. No.	Algorithm	Model parameters	Accuracy (10-fold CV)	Accuracy (hold-out)
1	k-NN	$k = 24$	67.33%	53.57%
2	Random forest	$mtry = 109$, $ntree = 10001$	73.33%	82.14%
3	Linear SVM	$C = 1$	75.33%	67.86%
4	SVM with RBF kernel	$C = 128$, $\sigma = 1.907349 \times 10^{-5}$	76%	71.43%

Table 6. Results on 104 subjects data at 0 and nearest to 48 h (Experiment 3).

Sr. No.	Algorithm	Model parameters	Accuracy (10-fold CV)	Accuracy (hold-out)
1	k-NN	$k = 3$	78.79%	77.78%
2	Random forest	$mtry = 109$, $ntree = 10001$	81.99%	80%
3	Linear SVM	$C = 1$	77.39%	80%
4	SVM with RBF kernel	$C = 3$, $\sigma = 3.051758 \times 10^{-5}$	82.83%	77.78%

Table 7. Results on 104 subjects data at 0 and nearest to onset or average onset time (Experiment 4).

Sr. No.	Algorithm	Model parameters	Accuracy (10-fold CV)	Accuracy (hold-out)
1	k-NN	$k = 4$	78.1%	77.78%
2	Random forest	$mtry = 109$, $ntree = 10001$	84.26%	77.78%
3	Linear SVM	$C = 1$	81.77%	75.56%
4	SVM with RBF kernel	$C = 8$, $\sigma = 3.051758 \times 10^{-5}$	85.45%	75.56%

due to the significance of the *"Onset Time"* which shows that the blood samples collected at nearest to *"Onset Time"* is playing important role in discrimination of the infected samples from non-infected samples.

The highest accuracy obtained at nearest to 48 h is 82.83% and at nearest to *"Onset Time"* is 85.45% which gives a positive sign that the prediction of respiratory viral infection at the early stage is possible with considerable accuracy.

6   Biomarker Identification

In this section, we show the top 10 important genes which are experimentally found to be the most important ones for the progression of respiratory viral infection and play an important role in the discrimination of infected samples from non-infected ones (see Table 8).

Random Forest also calculates the overall importance score for every feature. We used the caret package which calculates the overall importance score and scales it in a range from 0 to 100 [10]. We extracted the 109 genes which have highest overall importance score (100 to 15.97 in descending order) and plotted them to find the cut-off threshold to come up with 10 most important genes which contribute significantly in the progression of the disease (see Fig. 5).

Moreover, we compared top 109 genes selected using random forest at nearest to 48 h with top 109 genes selected at nearest to onset time and we found that

Table 8. The 10 most important genes with their overall importance score.

Sr. No.	Probe IDs	Gene symbol	Overall importance score
1	3434_at	IFIT1	100
2	23586_at	DDX58	92.9190292
3	5359_at	PLSCR1	77.908644
4	51056_at	LAP3	76.5473908
5	9111_at	NMI	74.5011703
6	23424_at	TDRD7	67.0779044
7	8743_at	TNFSF10	60.7319657
8	2633_at	GBP1	58.8176266
9	24138_at	IFIT5	53.9912712
10	4599_at	MX1	53.3913318

these top 10 genes shown in Fig. 5 are common in both categories which shows that these top 10 genes are significantly important at both the early timestamps, i.e., nearest to 48 h and nearest to onset time.

The Five-number summary of gene expression of input data for the identified top 10 genes at different timestamps can be seen in the form of boxplots shown in Fig. 6. The boxplots of the top 10 genes at timestamp 0, 48 and "*Onset Time*" support our findings. First, the boxplots support our claim that the top 10 genes reported by us are differentially expressed genes and contributing in

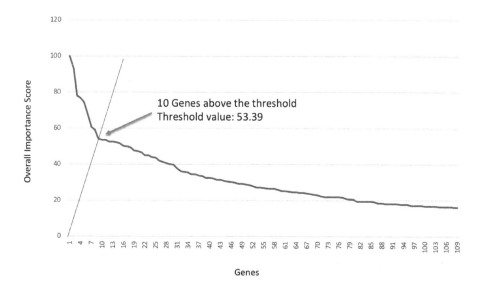

Fig. 5. Plot of overall importance score. 10 genes are above the cut-off threshold which are significantly the most important ones.

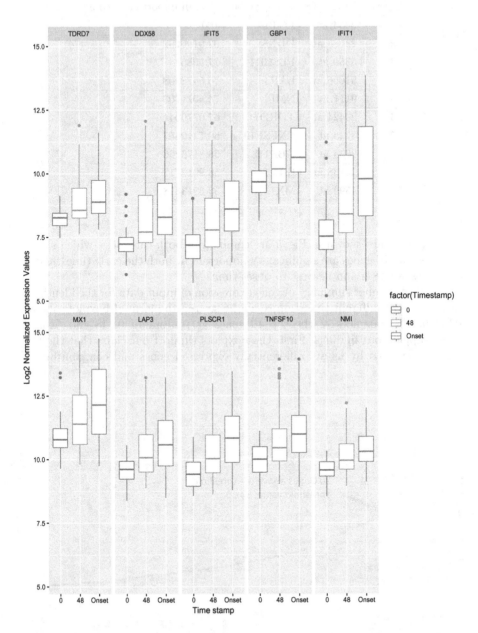

Fig. 6. Boxplots of the identified top 10 genes at 0 h, 48 h and *"Onset Time"*.

progression of respiratory viral infection as their median value of gene expression at 0 hours and at "*Onset Time*" has a significant difference. Second, the boxplots also support the importance of genes, for example, gene IFIT1 has the highest importance score which can be seen in boxplot in terms of the highest difference in median gene expression values. Third, these plots also support our finding that the "*Onset Time*" is a better choice for learning the predictive models.

7   Discussion

We have identified 10 top genes from a set of 12,023 genes. To understand the mechanism associated with these genes we performed Gene Set Enrichment Analysis (GSEA) of these genes [19]. To further understand the association of retrieved disease mechanisms we performed Transcription Factor (TF) analysis [21]. During TF analysis we integrated TRANSFAC [24], BioGPS [26] and JASPER database [16] and ran GSEA. The GSEA yielded the 441 associations against ten input genes. To understand the process associated with retrieved TFs and ten seed genes we performed functional annotation considering neighbouring genes (Table 9) and later without considering neighbouring genes (Table 10).

Table 9. Functional annotation and Disease enrichment analysis (DEA) with neighbour genes using Gene Set Enrichment Analysis (GSEA).

Gene symbol	p-value	Geneset friends	Total friends	GO annotation
IFIT1	1.32×10^{-18}	10	721	Interferon-induced protein with tetratricopeptide repeats 1
DDX58	4.24×10^{-17}	10	1020	DEAD(Asp-Glu-Ala-Asp) box polypeptide 58
IFIT5	1.89×10^{-15}	10	1491	Interferon-induced protein with tetratricopeptide repeats 5
GBP1	1.89×10^{-15}	10	1491	Guanylate binding protein 1, interferon-inducible
MX1	1.52×10^{-14}	10	1837	Myxovirus (influenza virus) resistance 1, interferon-inducible protein p78 (mouse)
PLSCR1	1.99×10^{-14}	10	1837	Phospholipid scramblase 1
TNFSF10	3.99×10^{-13}	10	2547	Tumour necrosis factor (ligand) superfamily, member 10
LAP3	5.67×10^{-11}	9	2505	Leucine aminopeptidase 3
NMI	9.54×10^{-11}	9	2655	N-myc (and STAT) interactor
TDRD7	7.75×10^{-10}	8	2018	Tumour domain containing 7

Here Geneset friends column explain how many genes contributed from the seed gene to establish the outcome. The MX1 gene has been known for its relevance to the intervention in the influenza virus infectious disease and it is known as the antiviral protein 1 [1,22]. IFIT1, IFIT5 have interferon-induced protein with tetratricopeptide repeats 1 as annotation which indicates it's role in viral pathogenesis [6]. DDX58 is cytoplasmic viral RNA receptor, that is also known as DDX58 (DExD/H-box helicase 58). GBP1 induces infectious virus production in primary human macrophages [9,27]. PLSCR1 are responsible for Hepatitis B virus replication with in-vitro and in-vivo both. LAP3 has already been predicted as principal viral response factor for all samples in H3N2 [3]. NMI has been reported as viral infection with Respiratory Syncytial Virus (RSV) infection and neuromuscular impairment (NMI) [23]. TDRD7 is known as interferon's antiviral action and responsible for paramyxovirus replication [20].

To explore the effect of captured mechanism further we conducted a DEA using GSEA outcomes and as a result, most of the genes appear to be aligned against response to virus (83), immune response (467), innate immune response (105). GSEA, a measure to define the inhibition of a gene alongside its nearest neighbours and known interactions not only helped to understand the virology aspect of ten seed genes but also associated factors and genes. As we can observe from Table 10 most of the genes are involved in antiviral infection and their extended neighbours are against the response to the virus or process related to immune the body against the virus attack.

Table 10. Functional annotation without neighbour genes using Gene Set Enrichment Analysis (GSEA).

GO biological process	p-value	GSEA enriched GENES
GO:0009615: response to virus (83)	$e^{-48.85}$	IRF7; PLSCR1; MX2; MX1; EIF2AK2; STAT1; BST2; IFIH1; TRIM22; IRF9; IFI35; DDX58; ISG15; RSAD2
GO:0006955: immune response (467)	$e^{-42.41}$	IFITM2; TAP1; IFITM3; IFI35; TNFSF10; GBP1; IFI6; CXCL10; IFI44L; OASL; OAS3; TRIM22; OAS2; PSMB9; OAS1; CXCL11; DDX58; IFIH1; SP110; PLSCR1
GO:0045087: innate immune response (105)	$e^{-11.45}$	IFIH1; DDX58; MX1; MX2; SP110

This provides a strong domain associated validation for these genes where core gene works as antiviral, and neighbour and interaction genes are immune and protective markers. This etiological discriminant prediction model and identified predictors is a potentially useful tool in epidemiological studies and viral infections.

8   Future Work

We will be extending this work to establish identified genes for pathogen related infection. Findings could have diagnostic and prognostic implications by informing patient management and treatment choice at the point of care. Thus, further our efforts in this direction will establish the power of non-linear mathematical models to analyze complex biomedical datasets and highlight key pathways involved in pathogen-specific immune responses. The implemented classification methodology will support future database updates or largely integrated knowledge graphs to include new viral infection database to establish diagnostically strong biomarker with phenotype data, which will enrich the classifiers. The sets of identified genes can potentiate the improvement of the selectivity of non-invasive infection diagnostics. Currently, any type of viral data with labelled samples (i.e. case/control) can be used to discover small sets of biomarkers. In future we will also be focusing on the following aspects:

- Predictive performance assessed with an n-fold cross-validation scheme and simulation of a validation with unseen samples of multiple databases having integrated knowledge graphs (i.e. external validation).
- Biomarker extraction and inference of the predictive model by incorporating time series analysis performed on the data that includes all the different time-points.
- Permutation test to statistically validate the predictive performance of the model. On this point, currently we have already achieved the following:
  - The variable importance represents the contribution of each biomarker at an early stage within the predictive model.
  - The variable direction indicates how the change in values affect the overall prediction (e.g. probability of the disease to occur).

9   Conclusions

In this work, we aim to use a hybrid approach that harnesses the power of both ML and database integration to provide new insights and improve understanding of viral etiology, particularly related to the mechanism of viral diseases. To achieve this we conducted four different experiments to assess the capability of ML algorithms to predict the state of disease at the early stage by analyzing gene expression data. We establish that the prediction at an early stage is possible with considerable accuracy, 82.83% accuracy at nearest to 48 h and 85.45% accuracy at nearest to onset time using 10-fold cross-validation, and accuracies of 80% and 82.14%, respectively on the hold-out test set. We got highest 10-fold cross-validation accuracy when all 104 subjects data are collected at 0 and nearest to onset or average onset time. This shows that for these kinds of studies if "*Onset Time*" is considered for learning the model then one can achieve considerably high accuracy in discrimination of infected from non-infected samples, however, it is observed that the accuracy on the hold-out test set is sometimes lower and sometimes higher than the 10-fold cross-validation accuracy, which means that the data has high variability and further analysis to capture this variability can improve the accuracy of prediction. The experiments indicate that the k-NN and linear SVM are not an ideal choice for these kinds of high dimensional datasets. By considering the fact that the Random Forest gives more stable and highest accuracy on unseen data (hold-out test set) for all the 4 experiments and due to its capability to assign importance score to variables, it is reasonable to choose Random Forest rather than the others. Moreover, we have identified top 10 most important genes which are having the maximum contribution in the progression of the respiratory viral infection at the early stage. The diagnosis and prevention of the respiratory viral infection at the early stage by targeting these genes can potentially improve the results than targeting the genes affected at the later stage of the infection.

Acknowledgements. This publication has emanated from research conducted with the financial support of Science Foundation Ireland (SFI) under Grant Number SFI/12/RC/2289, co-funded by the European Regional Development Fund.

References

1. Braun, B.A., Marcovitz, A., Camp, J.G., Jia, R., Bejerano, G.: Mx1 and Mx2 key antiviral proteins are surprisingly lost in toothed whales. Proc. Nat. Acad. Sci. **112**(26), 8036–8040 (2015)
2. Burges, C.J.: A tutorial on support vector machines for pattern recognition. Data Min. Knowl. Disc. **2**(2), 121–167 (1998)
3. Chen, M., et al.: Predicting viral infection from high-dimensional biomarker trajectories. J. Am. Stat. Assoc. **106**(496), 1259–1279 (2011)
4. Cunningham, P., Delany, S.J.: k-nearest neighbour classifiers. Mult. Classif. Syst. **34**, 1–17 (2007)
5. Díaz-Uriarte, R., De Andres, S.A.: Gene selection and classification of microarray data using random forest. BMC Bioinform. **7**(1), 3 (2006)
6. Fensterl, V., Sen, G.C.: Interferon-induced ifit proteins: their role in viral pathogenesis. J. Virol. **89**, 2462–2468 (2014). https://doi.org/10.1128/JVI.02744-14
7. Guyon, I., Weston, J., Barnhill, S., Vapnik, V.: Gene selection for cancer classification using support vector machines. Mach. Learn. **46**(1), 389–422 (2002)
8. Hsu, C.W., Chang, C.C., Lin, C.J.: A practical guide to support vector classification (2010)
9. Krapp, C., et al.: Guanylate binding protein (GBP) 5 is an interferon-inducible inhibitor of HIV-1 infectivity. Cell Host Microbe **19**(4), 504–514 (2016)
10. Kuhn, M.: Building predictive models in r using the caret package. J. Stat. Softw. Artic. **28**(5), 1–26 (2008)
11. Liaw, A., Wiener, M.: Classification and regression by randomForest. R News **2**(3), 18–22 (2002). http://CRAN.R-project.org/doc/Rnews/
12. Liu, T.Y., et al.: An individualized predictor of health and disease using paired reference and target samples. BMC Bioinform. **17**(1), 47 (2016)
13. McCloskey, B., Dar, O., Zumla, A., Heymann, D.L.: Emerging infectious diseases and pandemic potential: status quo and reducing risk of global spread. Lancet Infect. Dis. **14**(10), 1001–1010 (2014)
14. Molinari, N.A.M., et al.: The annual impact of seasonal influenza in the US: measuring disease burden and costs. Vaccine **25**(27), 5086–5096 (2007)
15. R Core Team: R: A Language and Environment for Statistical Computing. R Foundation for Statistical Computing, Vienna, Austria (2013). http://www.R-project.org/
16. Sandelin, A., Alkema, W., Engström, P., Wasserman, W.W., Lenhard, B.: Jaspar: an open-access database for eukaryotic transcription factor binding profiles. Nucleic Acids Res. **32**(suppl-1), D91–D94 (2004)

17. Scholkopf, B., et al.: Comparing support vector machines with Gaussian kernels to radial basis function classifiers. IEEE Trans. Signal Process. **45**(11), 2758–2765 (1997)
18. Statistics, L.B., Breiman, L.: Random forests. Machine Learning **45**, 5–32 (2001)
19. Subramanian, A., et al.: Gene set enrichment analysis: a knowledge-based approach for interpreting genome-wide expression profiles. Proc. Nat. Acad. Sci. **102**(43), 15545–15550 (2005)
20. Subramanian, G., et al.: A new mechanism of interferon's antiviral action: induction of autophagy, essential for paramyxovirus replication, is inhibited by the interferon stimulated gene, tdrd7. PLoS pathog. **14**(1), e1006877 (2018)
21. Vaquerizas, J.M., Kummerfeld, S.K., Teichmann, S.A., Luscombe, N.M.: A census of human transcription factors: function, expression and evolution. Nat. Rev. Genet. **10**(4), 252 (2009)
22. Verhelst, J., Parthoens, E., Schepens, B., Fiers, W., Saelens, X.: Interferon-inducible protein Mx1 inhibits influenza virus by interfering with functional viral ribonucleoprotein complex assembly. J. Virol. **86**(24), 13445–13455 (2012)
23. Wilkesmann, A., et al.: Hospitalized children with respiratory syncytial virus infection and neuromuscular impairment face an increased risk of a complicated course. Pediatr. Infect. Dis. J. **26**(6), 485–491 (2007)
24. Wingender, E., et al.: Transfac: an integrated system for gene expression regulation. Nucleic Acids Res. **28**(1), 316–319 (2000)
25. Woods, C.W., et al.: A host transcriptional signature for presymptomatic detection of infection in humans exposed to influenza H1N1 or H3N2. PloS One **8**(1), e52198 (2013)
26. Wu, C., et al.: Biogps: an extensible and customizable portal for querying and organizing gene annotation resources. Genome Biol. **10**(11), R130 (2009)
27. Zhu, Z., et al.: Nonstructural protein 1 of influenza a virus interacts with human guanylate-binding protein 1 to antagonize antiviral activity. PloS One **8**(2), e55920 (2013)

Automated Coding of Medical Diagnostics from Free-Text: The Role of Parameters Optimization and Imbalanced Classes

Luiz Virginio[(✉)] and Julio Cesar dos Reis

University of Campinas, Campinas, São Paulo, Brazil
luiz.virginio.jr@gmail.com, jreis@ic.unicamp.br

Abstract. The extraction of codes from Electronic Health Records (EHR) data is an important task because extracted codes can be used for different purposes such as billing and reimbursement, quality control, epidemiological studies, and cohort identification for clinical trials. The codes are based on standardized vocabularies. Diagnostics, for example, are frequently coded using the International Classification of Diseases (ICD), which is a taxonomy of diagnosis codes organized in a hierarchical structure. Extracting codes from free-text medical notes in EHR such as the discharge summary requires the review of patient data searching for information that can be coded in a standardized manner. The manual human coding assignment is a complex and time-consuming process. The use of machine learning and natural language processing approaches have been receiving an increasing attention to automate the process of ICD coding. In this article, we investigate the use of Support Vector Machines (SVM) and the binary relevance method for multi-label classification in the task of automatic ICD coding from free-text discharge summaries. In particular, we explored the role of SVM parameters optimization and class weighting for addressing imbalanced class. Experiments conducted with the Medical Information Mart for Intensive Care III (MIMIC III) database reached 49.86% of f1-macro for the 100 most frequent diagnostics. Our findings indicated that optimization of SVM parameters and the use of class weighting can improve the effectiveness of the classifier.

Keywords: Automated ICD coding · Multi-label classification
Imbalanced classes

1 Introduction

The Electronic Health Records (EHRs) are becoming widely adopted in the healthcare industry [1]. EHR is a software solution used to register health information about patients, as well as to manage health organizations activities for medical billing and even population health management. The data entered in the EHR usually contain both structured data (patient demographics, laboratory results, vital signs, etc.) and unstructured data (free-text notes).

Most of the records in an EHR are textual documents such as progress notes and discharge summaries entered by health professionals who attended the patient.

© Springer Nature Switzerland AG 2019
S. Auer and M.-E. Vidal (Eds.): DILS 2018, LNBI 11371, pp. 122–134, 2019.
https://doi.org/10.1007/978-3-030-06016-9_12

Discharge summary is a free-text document that is recorded in the moment of patient discharge. It describes the main health information about a patient during his/her visit to a hospital and provides final diagnosis, main exams, medication, treatments, etc. These unstructured data inserted as free text have the advantage of giving greater autonomy to health professionals for registering clinical information, but it entails issues for automatic data analysis [2].

In this scenario, extracting codes from EHR based on terminologies and standard medical classifications is an important task because the codes can be used for different purposes such as billing and reimbursement, quality control, epidemiological studies, and cohort identification for clinical trials [3]. Diagnosis coding, for example, is used not only for reporting and reimbursement purposes (in US, for example), but for research applications such as tracking patients with sepsis [4].

Usually, several EHR records are encoded in a standardized way by terminologies such as the International Classification of Diseases (ICD)[1] which is a taxonomy of diagnostic codes organized in a hierarchical structure. ICD codes are organized in a rooted tree structure, with edges representing is-a relationships between parents and children codes. More specifically, the ICD-9 contains more than 14 thousand classification codes for diseases. Codes contain three to five digits, where the first three digits represent disease category and the remaining digits represent subdivisions. For example, the disease category "essential hypertension" has the code 401, while its subdivisions are 401.0 - Malignant essential hypertension, 401.1 - Benign essential hypertension, and 401.9 - Unspecified essential hypertension.

Extracting codes from EHR textual documents requires the review of patient data searching for information that can be coded in a standardized manner. For example, evaluate discharge summary to assign ICD codes. Trained professional coders review the information in the patient discharge summary and manually assign a set of ICD codes according to the patient conditions described in the document [5]. However, assigning diagnosis codes performed by human coders is a complex and time-consuming process. In practical settings, there are many patients and the insertion of data and coding process require software support to be further effective.

Several proposals have been conducted to attempt automating the ICD coding process (e.g., [3, 6, 7]). A study conducted by Dougherty et al. showed that an ICD coding process assisted by an auto-coding improved coder productivity by over 20% on inpatient documentation [8]. Therefore, an automated system can help medical coders in the task of ICD coding and, consequently, reduce costs. However, this task has been shown to be a very challenging problem, especially because of the large number of ICD codes and the complexity of medical free-text [9].

According to our literature review, several research challenges remain opened in this direction. Medical free-text is difficult to be handled by machine learning approaches because misspellings and not unstandardized abbreviations often compromise their quality [10]. Besides, automated ICD coding is characterized to present aspects that negatively affect effectiveness such as large labels set, class imbalance, inter-class correlations, and large feature sets [7]. Despite these challenges, machine

---

[1] http://www.who.int/classifications/icd/en/.

learning approaches for automated coding are very promising because the model is automatically created from training data, without the need of human intervention.

In this paper, we aim to construct a model based on machine learning approaches for automatic ICD coding from free-text discharge summaries. In particular, we investigate the role of SVM parameters optimization and class weighting for imbalanced class addressing. In a machine learning perspective, a free-text sample could be considered as an instance in which one or more ICD codes can be assigned. It means that ICD codes (labels) are not exclusive and, therefore, a discharge summary can be labeled as belonging to multiple disease classes. That scenario is known as a multi-label classification task. In this work, we address multi-label classification problems into several multi-class, where each sample belongs to a single class. The results presented in our experimental study have shown that considering parameter values searching and the use of class weighting can bring improvements to the automatic coding task.

This article is organized as follows: Sect. 2 presents the related work. Section 3 introduces our experimental design. Then, Sect. 4 reports on our obtained results and discusses the findings. Section 5 presents the final considerations.

2 Related Work

Two approaches are usually explored in automated coding task of medical text: (i) Information Retrieval (IR) of codes from a dictionary; and (ii) machine learning or rule-based Text Classification (TC). In the first approach, an IR system is used to allow professional coders to search for a set of one or more terms in a dictionary [11]. TC approaches have been receiving an increasing attention in the task of medical text coding.

Several studies have proposed models for ICD coding and their methods ranged from manual rules to online learning. The best results for classification accuracy have been achieved by rules-based systems [12] in which hand-crafted expert rules are created. Nevertheless, these methods may be very time-consuming due to the necessity of creating hand-craft expert rules for all ICD codes.

Machine learning approaches are very promising because the model is automatically created from training data, without the need of human intervention. A literature review conducted by Stanfill et al. [13] concluded that most of studies presenting reliable results are inserted in controlled settings, often using normalized data and keeping a limited scope. For example, Zhang et al. [14] used SVMs and achieved a F1 score of 86.6%. However, they used only radiology reports with limited ICD-9 codes.

Perotte et al. [5] proposed the use of a hierarchy-based Support Vector Machines (SVM) model in the task of automated diagnosis code classification. The tests were conducted over the Medical Information Mart for Intensive Care (MIMIC II) dataset. The authors considered two different approaches for predicting ICD-9 codes: Flat SVM and hierarchy-based SVM. The flat SVM treated each ICD-9 code independently of each other whereas hierarchy-based SVM leveraged the hierarchical nature of ICD-9 codes into its modeling. The best results achieved a F1 score of 39.5% with the hierarchy-based SVM.

Several theoretical studies on multi-label classification have indicated that effectively exploiting correlations between labels can benefit the multi-label classification effectiveness [10]. Subotin et al. [15] proposed a method in which a previous model is trained to estimate the conditional probability of one code being assigned to a document, given that it is known that another code has been assigned to the same document. After, an algorithm applies this model to the output of an existing statistical auto-coder to modify the confidence scores of the codes. They tested their model for ICD-10 procedure codes.

Kavuluro et al. [7] conducted experiments to evaluate supervised learning approaches to automatically assign ICD-9 codes in three different datasets. They used different problem transformation approaches with different feature selection, training data selection, classifier chaining, and label calibration approaches. For the larger dataset, they achieved F1-score of 0.57 for codes with at least 2% of representation (diagnostics that were present in at least 2% of the records). Over all codes (1231 codes), they obtained a F1-score of 0.47, even with 80% of these codes having less than 0.5% of representation. They concluded that datasets with different characteristics and different scale (size of the texts, number of distinct codes, etc.) warrant different learning approaches.

Scheurwegs et al. explored a distributional semantic model using word2vec skip-gram model to generalize over concepts and retrieve relations between them. Their approach automatically searched concepts on Unified Medical Language System (UMLS) Metathesaurus[2], an integration of biomedical terminologies, using the Meta-Map[3] tool to extract named entities and semantic predications from free text. The datasets they used are in Dutch and are derived from the clinical data warehouse at the Antwerp University Hospital. They concluded that concepts derived from raw clinical texts outperform a bag-of-words approach for ICD coding.

Berndorfer and Henriksson [16] explored various text representations and classification models for assigning ICD-9 codes to discharge summaries in Medical Information Mart for Intensive Care III (MIMIC III)[4] database. For text representation, they compared two approaches: shallow and deep. The shallow representation describes each document as a bag-of-words using Term Frequency - Inverse Document Frequency (TF-IDF), while the deep representation describes the documents as a TF-IDF-weighted sum of semantic vectors that were learned using Word2Vec. The author still tested a combination strategy, in which features from the two representations are concatenated. For classification models, Berndorfer and Henriksson explored the Flat SVM and hierarchical SVM. They concluded that the best results, with F1-score of 39.25%, was obtained by combining models built using different representations.

Shi et al. [17] used deep learning approaches to automatically assign ICD-9 codes to discharge summaries from MIMIC-III database. They achieved a F1-score of 53%. Their results were obtained by not using the entire discharge summary; their experiments only considered the sections of 'discharge diagnosis' and 'final diagnosis',

---

[2] https://www.nlm.nih.gov/research/umls/about_umls.html.

[3] https://metamap.nlm.nih.gov/.

[4] https://mimic.physionet.org/.

where the description of patient diagnoses is found. Therefore, such approach was very dependent on the specificities of the database and presents difficulties to be generalized.

To the best of our knowledge based on the literature review, most of studies did not perform optimization of machine learning parameters. The studies have chosen the parameter values of the algorithms arbitrarily according to our interpretation. In addition, most of studies did not use approaches to address the problem of imbalanced class.

3   Materials and Methods

In this section, we present the materials and methods we used in the development of this work. We present the database used for testing and the procedure performed for model construction.

3.1   Dataset

The dataset used to extract the corpus of discharge summaries and respective ICD codes was MIMIC III. The discharge summaries correspond to 53.423 hospital admissions for adult patients between 2001 and 2012. ICD-9 was used to assign diagnosis codes to discharge diagnoses.

MIMIC III repository contains 55.177 discharge summaries and 6.985 different diagnosis codes. Only the 100 most frequent diagnostics were considered in this work. Therefore, we selected discharge summaries that had at least one of the 100 most frequent codes, resulting in 53.018 discharge summaries.

The distribution of labels among the samples is strongly imbalanced. The top three ICD-9 codes are:

- Unspecified essential hypertension (401.9) – present in 37.5% of the records
- Congestive heart failure, unspecified (428.0) – present in 23.8% of the records
- Atrial fibrillation (427.31) – present in 23.4% of the records

The hundredth most frequent ICD-9 code is "personal history of malignant neoplasm of prostate" (V10.46), which is presented in only 2% of the discharge summaries.

3.2   Procedure

We defined a pipeline to perform the classification task aiming to detect the ICD codes from the discharge summaries. Figure 1 presents the involved stages: pre-processing, dataset splitting, feature extraction, parameter search, and creation of prediction model. The following subsection describe details of the conducted procedure.

Data Handling. We constructed our dataset extracting all discharge summaries and respective diagnosis list from MIMIC III database. Therefore, each record in the dataset consists in a discharge summary and its respective ICD-9 codes list, which is represented by a vector of 100 dimensions in which each dimension corresponds to an

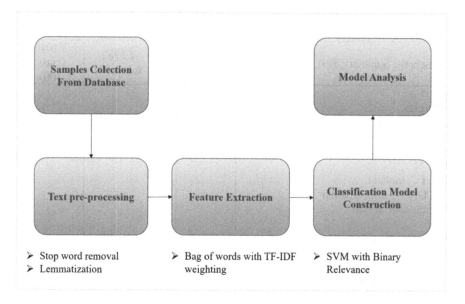

Fig. 1. Pipeline performed to construct the model for automated ICD coding

ICD-9 code. For a specific label in the record, if the corresponding ICD-9 code appears in the discharge summary diagnoses list, then its value in the vector is one, otherwise is zero.

For illustration purpose, Table 1 presents a sample of a record from the database. The first column represents the free-text of a discharge summary. The remaining columns represent each diagnosis code (class), where the column value is 1 or 0, depending whether the respective diagnosis was encoded for that discharge summary or not.

Dataset Splitting, Pre-processing, and Feature Extraction. Out of 53.018 discharge summaries, 80% were used for training and 20% for testing. The definition of the sets was performed in a stratified manner to maintain the proportion of classes in both sets. The training set was then used to define a vocabulary of tokens. Before tokenization, we implemented pre-processing actions expecting to improve the quality of classification and to reduce the index size of the training set. The following pre-processing tasks were performed: stop word removal, lemmatization, number removal, and special characters removal.

In stop word removal task, words that occur commonly across all the documents in the corpus are removed instead of being considered as a token. Generally, articles and pronouns are considered as stop words because they are not very discriminative. lemmatization which consists in a linguistic normalization. The variant forms of a term are reduced to a common form (lemma). The lemmatization process acts removing prefixes or suffixes of a term, or even transforming a verb to its infinitive form [18]. For stop word removal, we used the stop word removal function of the feature extraction

Table 1. Sample of a record in the dataset

Discharge Summary Text	4019 (class 1)	4280 (class 2)	...	E8782 (class 90)	V1046 (class 100)

[...]
Allergies:
    Amlodipine
 	
Attending:[**First   (LF) 898**]
Chief Complaint:
    COPD   exacerbation   / Shortness of Breath
 	
Major Surgical or Invasive Procedure:
    Intubation
    arterial line placement
    PICC line placement
    Esophagogastroduodenos-copy
 	
History of Present Illness:
87 yo F with h/o CHF, COPD on 5 L oxygen at baseline, tracheobronchomalacia   s/p stent, presents with acute dyspnea over several days, and lethargy. [...]
	1	0	...	0	0

module of the scikit-learn[5] library. For lemmatization, we used the class WordNetLemmatizer from Natural Language Toolkit (NLTK)[6] library.

The processed discharge summaries were then tokenized using unigram and bigram with TF-IDF weighting as features. The tokens with a document frequency strictly higher than 70% or lower than 1% were ignored resulting in 12.703 tokens. In this sense, we took the decision that the vocabulary as features does not contain too-frequent or too-rare unigrams and bigrams.

Parameters Searching and Prediction Model Creation. In this study, the classification task consisted in a multi-label classification in which one or more labels are assigned to a given record from the dataset. We used the Binary Relevance method to transform the multi-label problem into several binary classification problems. Therefore, we created one classifier per ICD-9 code.

---

[5] http://scikit-learn.org/stable/.

[6] https://www.nltk.org/.

We explored the SVM algorithm. SVM has important parameters like kernel, C, and gamma, which values have to be chosen for the training task. The majority of the studies found in literature for the code assignment problem, according to our knowledge, select parameters values arbitrarily. We assume that this decision might decrease the algorithm effectiveness. In this work, we performed a parameter search step, in which the training process was performed for each possible combination of predefined parameter values. The range of values for each parameter was defined as follows:

- Parameter kernel: [Linear, Radial Basis Function (RBF)]
- Parameter C: [0.02, 0.2, 1.0, 2.0, 4.0]
- Parameter gamma: [0.02, 0.2, 1.0, 2.0, 4.0]. Applicable only to the RBF kernel.

The parameter kernel specifies whether the SVM will perform a linear or a non-linear classification. To perform a linear classification, the kernel should be 'linear' while performing a non-linear classification requires a non-linear kernel, such as RBF [19]. The parameter C is related to the size of the margin of the SVM hyperplane, where low values of C will result in a large margin and high values of C result in a small margin. The size of the margin is strongly related to misclassification, because the smaller the margin, the smaller the misclassification [19]. However, lower misclassification on training set does not implicate in lower misclassification on testing set. Therefore, a larger margin may result in a more generalized classifier. Gamma is a free parameter of the Gaussian function of the RBF kernel.

Due to the unbalance of classes, another important parameter we considered was class weight. With this parameter, it was possible to penalize mistakes on the minority class proportionally to how under-represented it is. The *initial weight* for a class was computed as $N/(2 \times M)$, in which N is the number of records and M refers to the number of records in the respective class. This formula is widely used to deal with imbalanced classes in classification problems, because the lower the number of samples in a particular class, the higher is the *initial weight*.

The *initial weight* might be not enough to obtain a good effectiveness for too imbalanced classes. Therefore, besides using the *initial weight* value, we also used two higher values. We used the following range for class weight parameter: [None, *initial weight*, *initial weight* + 2, *initial weight* + 4].

According to the number of parameters and respective range of values, it was necessary to perform 90 SVM trainings (15 for linear kernel and 75 for RBF kernel). Due to computational power limitations, the parameter searching was performed in a subset corresponding to 30% of samples of the training set (12.724 samples). That subset was split in a second training set (80%) and validation (20%) set. Figure 2 illustrates the dataset splitting process.

A SVM model was created for each parameter combination using the second training set. The analysis of the model was tested in the validation set through the calculation of f1-score. The parameter combination values with best results were then selected as parameter values in the creation of the prediction model. Once one model is created for each class, such values can be different for distinct classes.

After the study concerning the parameters, the creation of prediction model (for each class) was performed using the training set with 42414 records (80% of the 53018 discharge summaries). The effectiveness of the method was evaluated in the test set

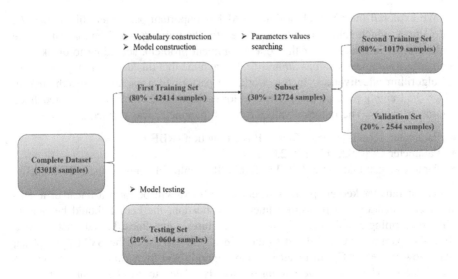

Fig. 2. Dataset splitting process

with 10.604 records. To this end, we explored the following evaluation metrics: recall, precision, and f1-score.

4   Results and Discussion

In this section, we present the results obtained with the construction of the classification model for ICD coding task. We highlight the influence of parameters optimization and the use of class weighting in the model construction.

4.1   Influence of Parameters and Class Weighting

After performing the searching for best combination of parameter values, we found that such values widely vary along the classes. The parameter 'C' varied between the following values: 1.0 (37 classes), 2.0 (31 classes), 0.2 (17 classes) and 4.0 (15 classes). For the 'gamma' parameter (applicable only to the RBF kernel), most classes presented the best results with the value 0.2 (52 classes), whereas five classes presented a value of 1.0 and three classes presented a value of 0.02.

For the 'kernel' parameter, 40 classes presented best results with a linear kernel, whereas 60 classes achieved better results with the RBF kernel. These results indicated the relevance of considering the RBF kernel. Usually, most studies in literature for ICD coding has approached the problem only using the linear kernel.

We addressed the problem of imbalanced classes with the use of class weighting. From 100 class in total, only two classes performed better without the need of using class weighting. These classes correspond to the diagnostics 276.8 – "Hypopotassemia" and 769 – "Respiratory distress syndrome in newborn" in ICD-9. The remaining 98 classes presented best results with the use of class weighting, highlighting the relevance of considering the class weighting as an approach to address the imbalanced class problem. According to the authors' knowledge, no other study has used this approach in literature for the studied problem.

4.2 Classifier Effectiveness

As previously mentioned, we tested the effectiveness of each model using the testing set. Table 2 summarizes the obtained results. We reached 49.86% for the f1-macro metric, which represents the mean of f1-score for all classes. The mean for recall score was 68.61% and the mean for precision score was 41.94%.

Table 3 presents the five classes with worst f1-score whereas Table 4 presents the five classes with best f1-score. The column "frequency index" in Tables 3 and 4 represents the position of the diagnosis in the database. For example, the diagnosis 42731 – "Atrial fibrillation" is the third most frequent diagnosis, whereas 99591 – "Sepsis" is the 92nd most frequent diagnosis. The higher the frequency index value, the lower the frequency of this diagnosis and, therefore, the more imbalanced is the respective class.

Results indicated that the classes presenting the worst effectiveness correspond to the most imbalanced classes. This suggests that the more diagnoses we consider, the lower the effectiveness of the model (cf. Fig. 3). For example, if we consider only the first 20 most frequent diagnostics, we obtain 65.43% of f1-macro against 49.86% if we consider the 100 most frequent diagnostics.

Table 2. Results summary

	Precision	Recall	F1-macro
Value	41.94%	68.61%	49.86%
Standard deviation	19.94%	14.67%	18.64%

Table 3. Five worst results and their respective classes

Diagnosis	Frequency index	Precision	Recall	F1-macro
E8788 - Other specified surgical operations and procedures causing abnormal patient reaction, or later complication, without mention of misadventure at time of operation	84	8.23%	76.40%	14.86%
27652 - Hypovolemia	81	9.91%	56.55%	16.87%
E8798 - Other specified procedures as the cause of abnormal reaction of patient, or of later complication, without mention of misadventure at time of procedure	69	13.22%	42.22%	20.14%
99591 - Sepsis	92	12.41%	68.85%	21.03%
2930 - Delirium due to conditions classified elsewhere	73	13.73%	65.14%	22.68%

Table 4. Five best results and their respective classes

Diagnosis	Frequency index	Precision	Recall	F1-macro
42731 - Atrial fibrillation	3	84.49%	88.86%	86.62%
V3000 - Single liveborn, born in hospital, delivered without mention of cesarean section	24	83.94%	89.26%	85.52%
7742 - Neonatal jaundice associated with preterm delivery	48	76.17%	97.98%	85.71%
V3001 - Single liveborn, born in hospital, delivered by cesarean section	36	81.05%	90.31%	85.43%
V290 - Observation for suspected infectious condition	13	76.33%	93.80%	84.17%

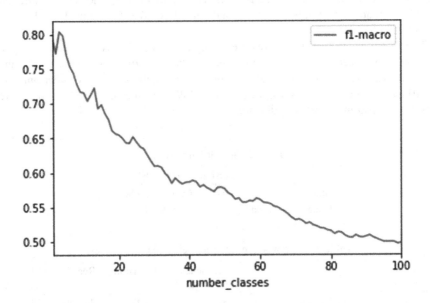

Fig. 3. Variation of f1-macro in relation to the number of classes

5   Conclusion

In this work, we constructed a model based on machine learning approaches for the task of automated ICD coding from free-text discharge summaries. The results we obtained highlight the importance of optimization of parameter as well as the use of class weighting approach to deal with imbalanced class problem.

We also highlight some limitations of this work. The computational power restrictions limited the range of parameters values to test as well as the number of samples in the second training set used for parameter optimization. We considered only the 100 most frequent diagnostics out of 6,985 diagnostics present in the database. Therefore, the most imbalanced classes (the less frequent diagnosis) were not considered. However, it is important to note that 96.6% of the diagnostics were assigned to only 1% or less of the discharge summaries.

Another important limitation is related specifically to the database characteristics. Most of the free-text discharge summaries present misspelling and abbreviations, which may have impaired the model effectiveness. In addition, the process of manually coding itself may have errors, which may have led to incorrect or incomplete list of diagnostics.

Acknowledgements. This work is supported by the São Paulo Research Foundation (FAPESP) (Grant #2017/02325-5)[7].

References

1. Chaudhry, B.: Systematic review: impact of health information technology on quality, efficiency, and costs of medical care. Ann. Intern. Med. **144**(10), 742 (2006)
2. Navas, H., Osornio, A.L., Baum, A., Gomez, A., Luna, D., de Quiros, F.G.B.: Creation and evaluation of a terminology server for the interactive coding of discharge summaries. Stud. Health Technol. Inform. **129**, 650–654 (2007)
3. Rios, A., Kavuluru, R.: Supervised extraction of diagnosis codes from EMRs: role of feature selection, data selection, and probabilistic thresholding. In: 2013 IEEE International Conference on Healthcare Informatics, pp. 66–73 (2013)
4. Scheurwegs, E., Luyckx, K., Luyten, L., Daelemans, W., Van den Bulcke, T.: Data integration of structured and unstructured sources for assigning clinical codes to patient stays. J. Am. Med. Inform. Assoc. **23**(e1), 11–19 (2016)
5. Perotte, A., Pivovarov, R., Natarajan, K., Weiskopf, N., Wood, F., Elhadad, N.: Diagnosis code assignment: models and evaluation metrics. J. Am. Med. Inform. Assoc. **21**(2), 231–237 (2014)
6. Hochreiter, S., Schmidhuber, J.: Long short-term memory. Neural Comput. **9**(8), 1735–1780 (1997)
7. Kavuluru, R., Rios, A., Lu, Y.: An empirical evaluation of supervised learning approaches in assigning diagnosis codes to electronic medical records. Artif. Intell. Med. **65**(2), 155–166 (2015)
8. Dougherty, M., Seabold, S., White, S.: Study Reveals hard facts on CAC. J. AHIMA **84**(7), 54–56 (2013)
9. Helwe, C., Elbassuoni, S., Geha, M., Hitti, E., Makhlouf Obermeyer, C.: CCS coding of discharge diagnoses via deep neural networks. In: Proceedings of the 2017 International Conference on Digital Health, DH 2017, pp. 175–179 (2017)
10. Wang, S., Chang, X., Li, X., Long, G., Yao, L., Sheng, Q.: Diagnosis code assignment using sparsity-based disease correlation embedding. IEEE Trans. Knowl. Data Eng. **28**(12), 3191–3202 (2016)
11. Rizzo, S.G., Montesi, D., Fabbri, A., Marchesini, G.: ICD code retrieval: novel approach for assisted disease classification. In: Ashish, N., Ambite, J.-L. (eds.) DILS 2015. LNCS, vol. 9162, pp. 147–161. Springer, Cham (2015). https://doi.org/10.1007/978-3-319-21843-4_12

---

[7] The opinions expressed in this work do not necessarily reflect those of the funding agencies.

12. Farkas, R., Szarvas, G.: Automatic construction of rule-based ICD-9-CM coding systems. BMC Bioinf. **9**(Suppl. 3), S10 (2008)
13. Stanfill, M.H., Williams, M., Fenton, S.H., Jenders, R.A., Hersh, W.R.: A systematic literature review of automated clinical coding and classification systems. J. Am. Med. Inform. Assoc. **17**(6), 646–651 (2010)
14. Zhang, Y.: A hierarchical approach to encoding medical concepts for clinical notes. In: Proceedings of the 46th Annual Meeting of the Association for Computational Linguistics on Human Language Technologies Student Research Workshop, HLT 2008, p. 67 (2008)
15. Subotin, M., Davis, A.R.: A method for modeling co-occurrence propensity of clinical codes with application to ICD-10-PCS auto-coding. J. Am. Med. Inform. Assoc. **23**(5), 866–871 (2016)
16. Berndorfer, S., Henriksson, A.: Automated diagnosis coding with combined text representations. Stud. Health Technol. Inform. **235**, 201–205 (2017)
17. Shi, H., Xie, P., Hu, Z., Zhang, M., Xing, E.P.: Towards automated ICD coding using deep learning, pp. 1–11 (2017)
18. Porter, M.F.: An algorithm for suffix stripping. Program **14**(3), 130–137 (1980)
19. Haykin, S.: Neural Networks and Learning Machines, vol. 3. Pearson, Upper Saddle River (2009)

A Learning-Based Approach to Combine Medical Annotation Results
(Short Paper)

Victor Christen[1(✉)], Ying-Chi Lin[1], Anika Groß[1], Silvio Domingos Cardoso[2,3],
Cédric Pruski[2], Marcos Da Silveira[2], and Erhard Rahm[1]

[1] University of Leipzig, Leipzig, Germany
{christen,lin,gross,rahm}@informatik.uni-leipzig.de
[2] LIST, Luxembourg Institute of Science and Technology,
Esch-sur-Alzette, Luxembourg
{silvio.cardoso,cedric.pruski,marcos.dasilveira}@list.lu
[3] LRI, University of Paris-Sud XI, Gif-sur-Yvette, France

Abstract. There exist many tools to annotate mentions of medical entities in documents with concepts from biomedical ontologies. To improve the overall quality of the annotation process, we propose the use of machine learning to combine the results of different annotation tools. We comparatively evaluate the results of the machine-learning based approach with the results of the single tools and a simpler set-based result combination.

Keywords: Biomedical annotation · Annotation tool
Machine learning

1 Introduction

The annotation of entities with concepts from standardized terminologies and ontologies is of high importance in the life sciences to enhance semantic interoperability and data analysis. For instance, exchanging and analyzing the results from different clinical trials can lead to new insights for diagnosis or treatment of diseases. In the healthcare sector there is an increasing number of documents such as electronic health records (EHRs), case report forms (CRFs) and scientific publications, for which a semantic annotation is helpful to achieve an improved retrieval of relevant observations and findings [1, 2].

Unfortunately, most medical documents are not yet annotated, e.g., as reported in [9] for CRFs, despite the existence of several tools to semi-automatically determine annotations. This is because annotating medical documents is highly challenging since documents may contain mentions of numerous medical entities that are described in typically large ontologies such as the Unified Medical Language System (UMLS) Metathesaurus. The mentions may also be ambiguous and incomplete and thus difficult to find within the ontologies. The

© Springer Nature Switzerland AG 2019
S. Auer and M.-E. Vidal (Eds.): DILS 2018, LNBI 11371, pp. 135–143, 2019.
https://doi.org/10.1007/978-3-030-06016-9_13

tools thus typically can find only a fraction of correct annotations and may also propose wrong annotations. Furthermore, the tools typically come with many configuration parameters making it difficult to use them in the best way.

Given the limitations of individual tools it is promising to apply several tools and to combine their results to improve overall annotation quality. In our previous work [11], we investigated already simple approaches to combine the results of three annotation tools based on set operations such as union, intersection and majority consensus. In this short paper, we propose and evaluate a machine learning (ML) approach for combining several annotation results.

Specifically, we make the following contributions:

- We propose a ML approach for combining the results of different annotation tools in order to improve overall annotation quality. It utilizes training data in the form of a so-called annotation vectors summarizing the scores of the considered tools for selected annotation candidates.
- We evaluate the new approach with different parameter and training settings and compare it with the results of single tools and the previously proposed combinations using set operations.

We first discuss related work on finding annotations and combining different annotation results. In Sect. 3, we propose the ML-based method. We then describe the evaluation methodology and analyze the results in Sect. 4. Finally, we conclude.

2   Related Work

Many annotation tools utilize a dictionary to store the concepts of the ontologies of interest (e.g., UMLS) to speedup the search for the most similar concepts for certain words of a document to annotate. Such dictionary-based tools include MetaMap, NCBO Annotator [8], IndexFinder [15], ConceptMapper [13], NOBLE Coder [14] cTAKES [12] and our own AnnoMap approach [7] that combines several string similarities and applies a post-processing to select the most promising annotations. There have also been annotation approaches using machine learning [4]. They can achieve good results but incur a substantial effort to provide suitable training data.

In our previous work [11], we combined annotation results for CRFs determined by the tools MetaMap, cTAKES and AnnoMap using the set-based approaches *union, intersection* and *majority*. The *union* approach includes the annotations from any tool to improve recall while *intersection* only preserves annotations found by all tools for improved precision. The *majority* approach includes the annotations found by a majority of tools, e.g., by at least two of three tools. Overall the set-based approach could significantly improve annotation quality, in particular for *intersection* and *majority*.

Though ML approaches have been used for annotating entities, so far they have rarely been applied for combining annotation results as we propose in this paper. Campos et al. utilized Conditional Random Fields model to recognize named entities of gene/protein terms using the results from three dictionary-based systems and one machine learning approach [5]. The learned combination could outperform combinations based on *union* or *intersection*. Our ML-based combination approach is inspired by methods proposed in record-linkage domain where the goal is to identify record pairs representing the same real-world entity [10]. Instead of a manually configured combination of different similarity values for different record attributes the ML approaches learn a classification model (e.g., using decision tree or SVM learning) based on a training set of matches and non-matches. The learned models automatically combine the individual similarities to derive at a match or non-match decision for every pair of records.

3   Machine Learning-Based Combination Approach

The task of *annotation* has as input a set of documents $D = \{d_1, d_2, \ldots, d_n\}$ to annotate, e.g., EHRs, CRFs or publications, as well as the ontology ON from which the concepts for annotation are to be found. The goal is to determine for each document fragment df (e.g., sentences) the set of its most precisely describing ontology concepts. The annotation result includes all associations between a document fragment df_j and its annotating concepts from ON. The problem we address is the *combination of multiple annotation results* for documents D and ontology ON that are determined by different tools. The tool-specific annotation results are obtained with a specific parameter configuration selected from a typically large number of possible parameter settings. The goal is to utilize complementary knowledge represented in the different input results to improve the overall annotation result, i.e., to find more correct annotations (better recall) and to reduce the number of wrongly proposed annotations (better precision).

The main idea of the proposed ML-based method is to train a classification model that determines whether an annotation candidate (df_j, c) between a document fragment df_j and a possibly annotating concept c is correct or not. The classification model is learned based on a set of positive and negative annotation examples for each tool (configuration). For each training example (df_j, c) we maintain a so-called annotation vector $\vec{av}$ with $n + 1$ elements, namely a quality score for each of the n annotation tools plus a so-called *basic score*. The basic score is a similarity between df_j and c that is independently computed from the annotation tools, e.g., based on a common string similarity function such as soft-TF/IDF or q-gram similarity. The use of the basic similarity is motivated by the observation that many concepts may be determined by only one or few tools leading to sparsely filled annotation vectors and thus little input for training the classification model. The learned classification model receives as input annotation vectors of candidate annotations and determines a decision whether the annotation is considered correct or not.

Figure 1 shows sample annotation vectors for three tools and the annotation of document fragment df_1. The table on the left shows the annotations found by the tools together with their scores (normalized to a value between 0 and 1). In total, the tools identify five different concepts resulting into the five annotation

	Identified annotations					
	Tool₁		Tool₂		Tool₃	
	concept	score	concept	score	concept	score
Document fragment df₁	C478762	1	C134877	0.3	C179926	0.86
	C134877	0.75	C179926	0.6	C243556	0.96
			C420838	0.3		

Annotation vectors				
Tool	Tool₁	Tool₂	Tool₃	Basic score
$\vec{av}_{(df1,C478762)}$	1	0	0	0.7
$\vec{av}_{(df1,C134877)}$	0.75	0.3	0	0.72
$\vec{av}_{(df1,C420838)}$	0	0.3	0	0.65
$\vec{av}_{(df1,C243556)}$	0	0	0.96	0.8
$\vec{av}_{(df1,C179926)}$	0	0.6	0.86	0.75

Fig. 1. Sample annotations and corresponding annotation vectors

vectors shown on the right of Fig. 1. For example, the annotation of df_1 with concept C478762 has the annotation vector $\vec{av}_{(df1,C478762)}$ of $(1, 0, 0, 0.7)$ since tool 1 identified this annotation with a score of 1, tools 2 and 3 did not determine this annotation (indicated by score 0), and the basic score is 0.7.

We use three classifiers: decision tree, random forest and support vector machines (SVM), to train classification models. A *decision tree* consists of nodes and each node represents a binary decision function based on a score threshold of a tool, e.g. $score_{MetaMap} > 0.7$. When an annotation vector $\vec{av}$ is input into a decision tree, decisions are made from the root node to the leaf node according to the values of $\vec{av}$. As output, $\vec{av}$ is classified as a correct or incorrect annotation. Random forest [3] utilizes an ensemble of decision trees and derives the classification decision from the most voted class of the individual decision trees. To determine a random forest classification model, each decision tree is trained by different samples of the training dataset. The goal of an *SVM* is to compute a hyperplane that separates the correct annotation vectors (represents a true annotation) from the incorrect ones. To separate vectors that are not linearly separable, SVM utilizes a kernel function to map the original vectors to a higher dimension so that the vectors can be separated.

A key step for the ML-based combination approach is the provision of suitable training data of a certain size. For this purpose, we determine annotation results with different tools and a specific configuration for a set of training documents. From the results we randomly select a subset of n annotations and generate the corresponding annotation vectors AV_{train} and label them as either correct or incorrect annotations. Providing a sufficient number of positive and negative training examples is of high importance to determine a classification model with enough discriminative power to correctly classify annotation candidates. To control the ratio between these two kinds of annotations we follow the approach of [10] and use a parameter $tpRatio$ (true positive ratio). For instance, $tpRatio = 0.4$ means 40% of all annotations in AV_{train} are correct. In our evaluation, we will consider the influence of both the training size n and $tpRatio$.

4   Evaluation and Results

We now evaluate our ML-based combination approach and compare it with the simpler set-based combination of annotation results. After the description of the experimental setup we analyze the influence of different training configurations and learners. In Sect. 4.3, we compare the results of the ML approach with the single tools and set-based combination. The evaluation focuses on the standard metrics *recall*, *precision* and their harmonic mean *F-measure* as the main indicator for annotation quality.

4.1   Experimental Setup

We use two datasets with medical forms (CRFs) for which a reference mapping exists: a dataset with forms on *eligibility criteria* (EC) and a dataset with *quality assurance* (QA) forms. The EC dataset contains 25 forms with 310 manually annotated questions. The QA dataset has 24 standardized forms with 543 annotated questions used in cardio-vascular procedures. The number of annotations in the reference mappings is 541 for EC and 589 for QA. These datasets have also been used in previous annotation evaluations [6,7] and turned out to be very challenging. For annotation we use five UMLS ontologies of version 2014AB: UMLS Metathesaurus, NCI Thesaurus, MedDRA, OAC-CHV, and SNOMED-CT_US. Since we use different subsets of UMLS in this paper and in the previous studies [7], the results are not directly comparable.

As in our previous study [11] we combine annotation results of the tools MetaMap, cTAKES and AnnoMap and apply the same set of configurations. In the annotation vectors, we use the normalized scores of the tools and determine the *basic score* by using soft-TF/IDF. For the classifiers (decision tree, random forest, SVM) we apply Weka as machine learning library. We generate training data of sizes 50, 100 or 200 selected from the *union* of the three tools. A *tpRatio* $\in \{0.2, 0.3, 0.4, 0.5\}$ is applied for each sample generation. For each ML test configuration (i.e., choice of classifier, sample size, *tpRatio* and tool configuration) we produce three randomly selected training sets and use each to generate a classifier model so that our results are not biased by just one sample. For each test configuration we measure average precision, average recall and macro F-measure that is based on the average precision and the average recall.

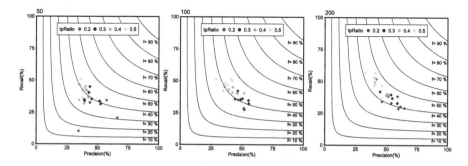

Fig. 2. Precision/recall results for different *tpRatio* values and training sizes n (dataset EC, random forest learning)

4.2   Machine Learning-Based Combination of Annotation Tools

For the analysis of our ML-based combination approach we first focus on the impact of parameter *tpRatio* and the size of the training sets. We then compare the three classifiers decision tree, random forest and SVM. Due to space restrictions we present only a representative subset of the results.

Figure 2 shows the annotation quality for dataset EC using random forest learning for different *tpRatios* (0.2 to 0.5) and three different training sizes (50, 100 and 200). Each data point represents the classification quality according to a certain *tpRatio* with a certain configuration of the considered tools. We observe that data points with the same *tpRatios* are mostly grouped together indicating that this parameter is more significant than other configuration details. We further observe for all training sizes that models trained with a larger *tpRatios* of 0.5 or 0.4 tend to reach a higher recall (but lower precision) than for smaller *tpRatios* values. Apparently low *tpRatio* values provide too few correct annotations so that the learned models are not sufficiently able to classify correct annotations as correct. By contrast, higher *tpRatio* values can lead to models that classify more incorrect annotations as a correct thereby reducing precision. For random forest, a *tpRatio* of 0.4 is generally a good compromise setting.

Figure 2 also shows that larger training sizes tend to improve F-measure since the data points for the right-most figure (training size $n = 200$) are mostly above the F-measure line of 50% while this is not the case for the left-most figure ($n = 50$). Figure 3 reveals the influence of the training size in more detail by showing the macro-average precision, recall and F-measure obtained by random forest using different training sizes. For both datasets, EC and QA, we observe that larger training sizes help to improve both precision and recall and thus F-measure. Hence, average F-measure improved from 40.1% to 42.5% for dataset EC and even from 52.0% to 56.9% for QA when the training size increases from 50 to 200 annotation samples.

Figure 4 depicts the macro-average precision, recall and F-measure over different *tpRatios*, sample sizes and configurations. For both datasets, random forest obtains the best recall values (EC: 40.0%, QA: 46.8%) while decision tree

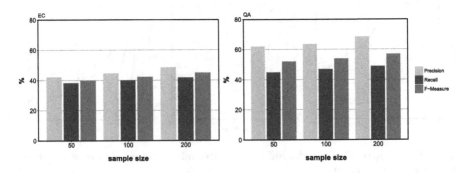

Fig. 3. Impact of training sizes on annotation quality for datasets EC and QA

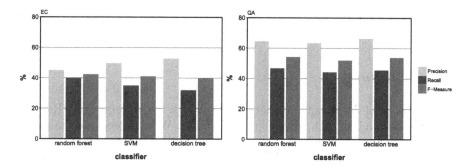

Fig. 4. Average annotation quality for random forest, SVM and decision tree.

achieves the best precision (EC: 52.9%, QA: 66.4%). In terms of average F-measure the three learning approaches are relatively close together, although random forest (42.4%) outperforms SVM and decision tree by 1.4% resp. 2.5% for EC. For the QA dataset, random forest (54.3%) outperforms decision tree and SVM by 0.3% resp. 2.2%. Moreover, we experimentally tested our approach with or without using the basic scores in addition to the tool results. We observed that using the basic score improves F-Measure by 1.6% (EC) and 1% (QA), indicating that it is valuable to improve annotation results.

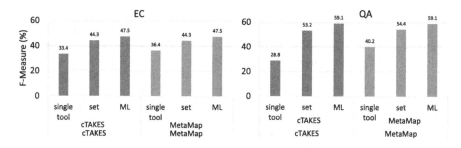

Fig. 5. Summarizing F-measure results for cTAKES and MetaMap and the set-based and ML-based result combinations for the EC and QA datasets.

4.3   Comparison with Set-Based Combination Approaches

We finally compare the annotation quality for the ML-based combinations with that of the individual tools cTAKES and MetaMap as well as with the results for the set-based combinations proposed in [11]. Figure 5 summarizes the best F-measure results for both datasets. We observe that the F-measure of the individual tools is substantially improved by both the set-based and ML-based combination approaches, especially for cTAKES (by about a factor 3–4.5). The ML-based combination outperforms the set-based combinations for both datasets. Consequently, the best results can be improved for EC (from 44.3% to 47.5%)

and QA (from 56.1% to 59.1%) by using a sample size of 200. This underlines the effectiveness of the proposed ML-based combination approach.

5   Conclusions

The annotation of documents in healthcare such as medical forms or EHRs with ontology concepts is of high benefit but challenging. We proposed and evaluated a machine learning approach to combine the annotation results of several tools. Our evaluation showed that the ML-based approach can dramatically improve the annotation quality of individual tools and that it also outperforms simpler set-based combination approaches. The evaluation showed that the improvements are already possible for small training sizes (50–200 positive and negative annotation examples) and that random forest performs slightly better than decision tree or SVM learning. In future work, we plan to apply the ML-based combination strategy to annotate further kinds of documents and to use machine learning also in the generation of annotation candidates.

References

1. TIES-Text Information Extraction System (2017). http://ties.dbmi.pitt.edu/
2. Abedi, V., Zand, R., Yeasin, M., Faisal, F.E.: An automated framework for hypotheses generation using literature. BioData Min. **5**(1), 13 (2012)
3. Breiman, L.: Random forests. Mach. Learn. **45**(1), 5–32 (2001)
4. Campos, D., Matos, S., Oliveira, J.: Current methodologies for biomedical named entity recognition. In: Biological Knowledge Discovery Handbook: Preprocessing, Mining, and Postprocessing of Biological Data, pp. 839–868 (2013)
5. Campos, D., et al.: Harmonization of gene/protein annotations: towards a gold standard MEDLINE. Bioinformatics **28**(9), 1253–1261 (2012)
6. Christen, V., Groß, A., Rahm, E.: A reuse-based annotation approach for medical documents. In: Groth, P., et al. (eds.) ISWC 2016. LNCS, vol. 9981, pp. 135–150. Springer, Cham (2016). https://doi.org/10.1007/978-3-319-46523-4_9
7. Christen, V., Groß, A., Varghese, J., Dugas, M., Rahm, E.: Annotating medical forms using UMLS. In: Ashish, N., Ambite, J.-L. (eds.) DILS 2015. LNCS, vol. 9162, pp. 55–69. Springer, Cham (2015). https://doi.org/10.1007/978-3-319-21843-4_5
8. Dai, M., et al.: An efficient solution for mapping free text to ontology terms. In: AMIA Summit on Translational Bioinformatics, vol. 21 (2008)
9. Dugas, M., et al.: Portal of medical data models: information infrastructure for medical research and healthcare. Database: J. Biol. Databases Curation (2016)
10. Köpcke, H., Thor, A., Rahm, E.: Learning-based approaches for matching web data entities. IEEE Internet Comput. **14**(4), 23–31 (2010)
11. Lin, Y.-C., et al.: Evaluating and improving annotation tools for medical forms. In: Da Silveira, M., Pruski, C., Schneider, R. (eds.) DILS 2017. LNCS, vol. 10649, pp. 1–16. Springer, Cham (2017). https://doi.org/10.1007/978-3-319-69751-2_1
12. Savova, G.K., et al.: Mayo clinical Text Analysis and Knowledge Extraction System (cTAKES): architecture, component evaluation and applications. JAMIA **17**(5), 507–513 (2010)

13. Tanenblatt, M.A., Coden, A., Sominsky, I.L.: The ConceptMapper approach to named entity recognition. In: Proceedings of LREC, pp. 546–551 (2010)
14. Tseytlin, E., Mitchell, K., Legowski, E., Corrigan, J., Chavan, G., Jacobson, R.S.: NOBLE-Flexible concept recognition for large-scale biomedical natural language processing. BMC Bioinform. **17**(1), 32 (2016)
15. Zou, Q., et al.: IndexFinder: a knowledge-based method for indexing clinical texts. In: Proceedings of AMIA Annual Symposium, pp. 763–767 (2003)

Knowledge Graph Completion to Predict Polypharmacy Side Effects

Brandon Malone[(✉)] , Alberto García-Durán, and Mathias Niepert

NEC Laboratories Europe, Kürfursten-Anlage 36, 69115 Heidelberg, Germany
{brandon.malone,alberto.duran,mathias.niepert}@neclab.eu

Abstract. The polypharmacy side effect prediction problem considers cases in which two drugs taken individually do not result in a particular side effect; however, when the two drugs are taken in combination, the side effect manifests. In this work, we demonstrate that multi-relational knowledge graph completion achieves state-of-the-art results on the polypharmacy side effect prediction problem. Empirical results show that our approach is particularly effective when the protein targets of the drugs are well-characterized. In contrast to prior work, our approach provides more interpretable predictions and hypotheses for wet lab validation.

Keywords: Knowledge graph · Embedding · Side effect prediction

1 Introduction

Disease and other health-related problems are often treated with medication. In many cases, though, multiple medications may be given to treat either a single condition or to account for co-morbidities. However, such combinations significantly increase the risk of unintended side effects due to unknown drug-drug interactions.

In this work, we show that multi-relational knowledge graph (KG) completion gives state-of-the-art performance in predicting these unknown drug-drug interactions. The KGs are multi-relational in the sense that they contain edges with different types. We formulate the problem as a multi-relational link prediction problem in a KG and adapt existing graph embedding strategies to predict the interactions. In contrast to prior approaches for the polypharmacy side effect problem, we incorporate interpretable features; thus, our approach naturally yields explainable predictions and suggests hypotheses for wet lab validation. Further, while we focus on the side effect prediction problem, our approach is general and can be applied to any multi-relational link prediction problem.

Much recent work has considered the problem of predicting drug-drug interactions (e.g. [2,13] and probabilistic approaches like [9]). However, these approaches only consider *whether* an interaction occurs; they do not consider the *type of interaction* as we do here. Thus, these methods are not directly

© Springer Nature Switzerland AG 2019
S. Auer and M.-E. Vidal (Eds.): DILS 2018, LNBI 11371, pp. 144–149, 2019.
https://doi.org/10.1007/978-3-030-06016-9_14

Table 1. Size statistics of the graph

	Count
Proteins	19 089
Drugs	645
Protein-protein interactions	715 612
Drug-drug interactions	4 649 441
Drug-protein target relationships	11 501
Mono side effects	174 977
Distinct mono side effects	10 184
Distinct polypharmacy side effects	963

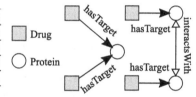

Fig. 1. Types of relational features.

comparable. The recently-proposed DECAGON approach [14] is most similar to ours; they also predict types of drug-drug interactions. However, they use a complicated combination of a graph convolutional network and a tensor factorization. In contrast, we use a neural KG embedding method in combination with a method to incorporate rule-based features. Hence, our method explicitly captures meaningful *relational features*. Empirically, we demonstrate that our method outperforms DECAGON in Sect. 4.

2 Datasets

We use the publicly-available, preprocessed version of the dataset used in [14].[1] It consists of a multi-relational knowledge graph with two main components: a protein-protein and a drug-drug interaction network. Known drug-protein target relationships connect these different components. The protein-protein interactions are derived from several existing sources; it is filtered to include only experimentally-validated physical interactions in human. The drug-drug interactions are extracted from the TWOSIDES database [11]. The drug-protein target relationships are experimentally-verified interactions from the STITCH [10] database. Finally, the SIDER [6] and OFFSIDES [11] databases were used to identify mono side effects of each drug. Please see Table 1 for detailed statistics of the size and density of each part of the graph. For more details, please see [14]. Each drug-drug link corresponds to a particular polypharmacy side effect. Our goal will be to predict missing drug-drug links.

3 Methods

KG embedding methods learn vector representations for entities and relation types of a KG [1]. We investigate the performance of DISTMULT [12], a commonly-used KG embedding method whose symmetry assumption is well-suited to this problem due to the symmetric nature of the drug-drug (polypharmacy side effect) relation type. The advantage of KG embedding methods are

---

[1] Available at http://snap.stanford.edu/decagon.

their efficiency and their ability to learn fine-grained entity types suitable for downstream tasks without hand-crafted rules. These embedding methods, however, are less interpretable than rule-based approaches and cannot incorporate domain knowledge.

A *relational feature* is a logical rule which is evaluated in the KG to determine its truth value. For instance, the formula $(\text{drug}_1, \text{hasTarget}, \text{protein}_1) \land$ $(\text{drug}_2, \text{hasTarget}, \text{protein}_1)$ corresponds to a binary feature which has value 1 if both drug_1 and drug_2 have protein_1 as a target, and 0 otherwise. In this work, we leverage relational features modeling drug targets with the relation type hasTarget and protein-protein interactions with the relation type interactsWith. Figure 1 depicts the two features types we use in our polypharmacy model. For a pair of entities (h, t), the relational feature vector is denoted by $r_{(h,t)}$. Relational features capture concrete relationships between entities; thus, as shown in Sect. 4, they offer explanations for our predictions.

KBLRN is a recently proposed framework for end-to-end learning of knowledge graph representations [4]. It learns a product of experts (PoE) [5] where each expert is responsible for one feature type. In the context of KG representation learning, the goal is to train a PoE that assigns high probability to true triples and low probabilities to triples assumed to be false. Let $d = (h, r, t)$ be a triple. The specific experts we use are defined as

$$f_{(r,L)}(d \mid \theta_{(r,L)}) = \begin{cases} \exp((e_h * e_t) \cdot w^r) \\ 1 \text{ for all } r' \neq r \end{cases} \text{ and } f_{(r,R)}(d \mid \theta_{(r,R)}) = \begin{cases} \exp\left(r_{(h,t)} \cdot w^r_{rel}\right) \\ 1 \text{ for all } r' \neq r \end{cases}$$

where $*$ is the element-wise product, $\cdot$ is the dot product, e_h and e_t are the embedding of the head and tail entity, respectively, and w^r, w^r_{rel} are the parameter vectors for the embedding and relational features for relation type r. The probability of triple $d = (h, r, t)$ is now

$$p(d \mid \theta) = \frac{f_{(r,L)}(d \mid \theta_{(r,L)}) \, f_{(r,R)}(d \mid \theta_{(r,R)})}{\sum_c f_{(r,L)}(c \mid \theta_{(r,L)}) \, f_{(r,R)}(c \mid \theta_{(r,R)})},$$

where c indexes all possible triples. As proposed in previous work, we approximate the gradient of the log-likelihood by performing negative sampling [4].

4   Experimental Results

We now empirically evaluate our proposed approach based on multi-relational knowledge graph completion to predict polypharmacy side effects.

Dataset Construction. We follow the common experimental design previously used [14] to construct our dataset. The knowledge graph only contains "positive" examples for which polypharmacy side effects exist. Thus, we create a set of negative examples by randomly selecting a pair of drugs and a polypharmacy side effect which does not exist in the knowledge graph. We ensure that the number of positive and negative examples of each polypharmacy side effect are equal. We then use stratified sampling to split the records in training, validation and testing sets.

We use an instance of the relational feature types depicted in Fig. 1 if it occurs at least 10 times in the KG. We choose these relational feature types because they offer a biological explanation for polypharmacy side effects; namely, a polypharmacy side effect may manifest due to unexpected combinations or interactions on the drug targets.

Baselines. We first compare our proposed approach to DECAGON [14]. Second, we consider each drug as a binary vector of indicators for each mono side effect and gene target. We construct training, validation and testing sets by concatenating the vectors of the pairs of drugs described above. We predict the likelihood of each polypharmacy side effect given the concatenated vectors.

Complete DECAGON *Dataset.* We first consider the same setting considered previously [14]. As shown in Table 2(top), our simple baseline, DISTMULT, and KBLRN all outperform DECAGON.

Drug-Drug Interactions Only. Next, we evaluate polypharmacy side effect prediction based solely on the pattern of other polypharmacy side effects. Specifically, we completely remove the drug-protein targets and protein-protein interactions from the KG; thus, we use only the drug-drug polypharmacy side effects in the training set for learning. We focus on DISTMULT and KBLRN since they outperformed the other methods in the first setting.

Surprisingly, the results in Table 2(middle) show that both DISTMULT and KBLRN perform roughly the same (or even *improve* slightly) in this setting, despite discarding presumably-valuable drug target information. However, as shown in Table 1, few drugs have annotated protein targets. Thus, we hypothesize that the learning algorithms ignore this information due to its sparsity.

Drugs with Protein Targets Only. To test this hypothesis, we remove all drugs which do not have any annotated protein targets from the KG (and the associated triples from the dataset). That is, the drug target information is no longer "sparse", in that all drugs in the resulting KG have protein targets.

The results in Table 2(bottom) paint a very different picture than before; KBLRN significantly outperforms DISTMULT. These results show that the combination of learned (or embedding) features and relational features can significantly improve performance when the relational features are present in the KG.

Explanations and Hypothesis Generation. The relational features allow us to explain predictions and generate new hypotheses for wet lab validation. We chose one of our high-likelihood predictions and "validated" it via literature evidence. In particular, the ranking of the drug combination CID115237 (paliperidone) and CID271 (calcium) for the side effect "pain" increased from 24 223 when using only the embedding features (of 58 029 pairs of drugs for which "pain" is not a known side effect) to a top-ranked pair when also using the relational features. Inspection of the relational features shows that the interaction

Table 2. The performance of each approach on the pre-defined test set. The measures are: area under the receiver operating characteristic curve (AuROC), area under the precision-recall curve (AuPR), and the average precision for the top 50 predictions for each polypharmacy side effect (AP@50). The best result within each group is in bold.

Method	AuROC	AuPR	AP@50
Baseline	0.896	0.859	0.812
DECAGON (values reported in [14])	0.872	0.832	0.803
DISTMULT	**0.923**	**0.898**	**0.899**
KBLRN	0.899	0.878	0.857
DISTMULT (drug-drug interactions only)	**0.931**	**0.909**	**0.919**
KBLRN (drug-drug interactions only)	0.894	0.886	0.892
DISTMULT (drugs with protein targets only)	0.534	0.545	0.394
KBLRN (drugs with protein targets only)	**0.829**	**0.797**	**0.774**

between lysophosphatidic acid receptor 1 (LPAR1) and matrix metallopeptidase 2 (MMP2) is particularly important for this prediction. The MMP family is known to be associated with inflammation (pain) [7]. Independently, calcium already upregulates MMP2 [8]. Paliperidone upregulates LPAR1, which in turn has been shown to promote MMP activiation [3]. Thus, palperidone indirectly exacerbates the up-regulation of MMP2 already caused by calcium; this, then, leads to increased pain. Hence, the literature confirms our prediction discovered due to the relational features.

5   Discussion

We have shown that multi-relational knowledge graph completion can achieve state-of-the-art performance on the polypharmacy side effect prediction problem. Further, relational features offer explanations for our predictions; they can then be validated via the literature or wetlab. In the future, we plan to extend this work by considering additional features of nodes in the graph, such as Gene Ontology annotations for the proteins and chemical structure of the drugs.

References

1. Bordes, A., Usunier, N., García-Durán, A., Weston, J., Yakhnenko, O.: Translating embeddings for modeling multi-relational data. In: Advances in Neural Information Processing Systems 26 (2013)
2. Cheng, F., Zhao, Z.: Machine learning-based prediction of drug-drug interactions by integrating drug phenotypic, therapeutic, chemical, and genomic properties. J. Am. Med. Inform. Assoc. **21**(e2), e278–e286 (2014)
3. Fishman, D.A., Liu, Y., Ellerbroek, S.M., Stack, M.S.: Lysophosphatidic acid promotes matrix metalloproteinase (MMP) activation and MMP-dependent invasion in ovarian cancer cells. Cancer Res. **61**(7), 3194–3199 (2001)

4. García-Durán, A., Niepert, M.: KBLRN: end-to-end learning of knowledge base representations with latent, relational, and numerical features. In: Proceedings of the 34th Conference on Uncertainty in Artificial Intelligence (2018)
5. Hinton, G.E.: Training products of experts by minimizing contrastive divergence. Neural Comput. **14**(8), 1771–1800 (2002)
6. Kuhn, M., Letunic, I., Jensen, L.J., Bork, P.: The SIDER database of drugs and side effects. Nucl. Acids Res. **44**(D1), D1075–D1079 (2016)
7. Manicone, A.M., McGuire, J.K.: Matrix metalloproteinases as modulators of inflammation. Semin. Cell Dev. Biol. **19**(1), 34–41 (2008)
8. Munshi, H.G., Wu, Y.I., Ariztia, E.V., Stack, M.S.: Calcium regulation of matrix metalloproteinase-mediated migration in oral squamous cell carcinoma cells. J. Biol. Chem. **277**(44), 41480–41488 (2002)
9. Sridhar, D., Fakhraei, S., Getoor, L.: A probabilistic approach for collective similarity-based drug-drug interaction prediction. Bioinformatics **32**(20), 3175–3182 (2016)
10. Szklarczyk, D., Santos, A., von Mering, C., Jensen, L.J., Bork, P., Kuhn, M.: STITCH 5: augmenting protein-checical interaction networks with tissue and affinity data. Nucl. Acids Res. **44**, D380–D384 (2016)
11. Tatonetti, N.P., Ye, P.P., Daneshjou, R., Altman, R.B.: Data-driven prediction of drug effects and interactions. Sci. Transl. Med. **4**(125), 125ra31 (2012)
12. Yang, B., tau Yih, S.W., He, X., Gao, J., Deng, L.: Embedding entities and relations for learning and inference in knowledge bases. In: Proceedings of the 3rd International Conference on Learning Representations (2015)
13. Zhang, W., Chen, Y., Liu, F., Luo, F., Tian, G., Li, X.: Predicting potential drug-drug interactions by integrating chemical, biological, phenotypic and network data. BMC Bioinform. **18**, 18 (2017)
14. Zitnik, M., Agrawal, M., Leskovec, J.: Modeling polypharmacy side effects with graph convolutional networks. Bioinformatics **34**(13), 457–466 (2018)

Big Biomedical Applications

Lung Cancer Concept Annotation
from Spanish Clinical Narratives

Marjan Najafabadipour[1] [ID], Juan Manuel Tuñas[1] [ID],
Alejandro Rodríguez-González[1,2(✉)] [ID], and Ernestina Menasalvas[1,2]

[1] Centro de Tecnología Biomédica, Universidad Politécnica de Madrid,
Madrid, Spain
{m.najafabadipour,alejandro.rg,
ernestina.menasalvas}@upm.es, juan.tunas@ctb.upm.es
[2] ETS de Ingenieros Informáticos, Universidad Politécnica de Madrid,
Madrid, Spain

Abstract. Recent rapid increase in the generation of clinical data and rapid development of computational science make us able to extract new insights from massive datasets in healthcare industry. Oncological Electronic Health Records (EHRs) are creating rich databases for documenting patient's history and they potentially contain a lot of patterns that can help in better management of the disease. However, these patterns are locked within free text (unstructured) portions of EHRs and consequence in limiting health professionals to extract useful information from them and to finally perform Query and Answering (Q&A) process in an accurate way. The Information Extraction (IE) process requires Natural Language Processing (NLP) techniques to assign semantics to these patterns. Therefore, in this paper, we analyze the design of annotators for specific lung cancer concepts that can be integrated over Apache Unstructured Information Management Architecture (UIMA) framework. In addition, we explain the details of generation and storage of annotation outcomes.

Keywords: Electronic health record · Natural language processing
Named entity recognition · Lung cancer

1 Introduction

Cancer is still one of the major public health issues, ranked with the second leading cause of death globally [1]. Across the Europe, lung cancer was estimated with 20.8% (over 266,000 persons) of all cancer deaths in 2011 [2] and the highest economic cost of 15% (18.8 billion) of overall cancer cost in 2009 [3]. Early diagnoses of cancer decreases its mortality rate [4]. Hence, a great attention on diagnoses is a key factor for both the effective control of the disease as well as the design of treatment plans.

Classically, the treatment decisions on lung cancer patients have been based upon histology of the tumor. According to World Health Organization (WHO), there are two broad histological subtypes of lung cancer: (1) Small Cell Lung Cancer (SCLC); and (2) Non-Small Cell Lung Cancer (NSCLC) [5]. NSCLC can be further defined at the molecular level by recurrent driver mutations [6] where mutations refer to any changes in the DNA sequence of a cell [7]. Tumor Mutations can occur in multiple oncogenes,

© Springer Nature Switzerland AG 2019
S. Auer and M.-E. Vidal (Eds.): DILS 2018, LNBI 11371, pp. 153–163, 2019.
https://doi.org/10.1007/978-3-030-06016-9_15

including in: Epidermal Growth Factor Receptor (EGFR), Anaplastic Lymphoma Kinase (ALK), and Ros1 proto-oncogene receptor tyrosine kinase [8]. These oncogenes are Receptor Tyrosine Kinases, which can activate pathways associated with cell growth and proliferation [9–11].

One of the preliminary diagnoses factor of a cancer is its tumor stage. This factor plays a significant role on making decisions for developing treatment plans. The American Joint Committee on Cancer (AJCC) manual [13] specifies two standard systems for measuring the cancer stage [14]: (1) stage grouping and (2) TNM. The stage grouping system encodes the tumor stages using roman numerals, whereas the TNM system makes use of three parameters: (1) the size of tumor (T); (2) the number of lymph nodes (N); and (3) the presence of Metastasis (M).

According to International Consortium for Health Outcomes Measurement (ICHOM), Performance Status (PS) is a strong individual predictor of survival in lung cancer. The ICHOM working group recommended measuring PS as part of diagnoses per the Eastern Cooperative Oncology Group (ECOG) [14]. In addition to ECOG, Karnofsky is another scale for measuring PS [15]. These scales are used by doctors and researchers to assess the progress of a patient's disease, the effects of the disease on daily and living abilities of a patient and to determine appropriate treatment and prognosis [16].

Towards the digitization of medical data, these data have been stored in computerized medical records, named EHRs. EHRs are rich clinical documents containing information about diagnoses, treatments, laboratory results, discharge summaries, to name a few, which can be used to support clinical decision support systems and allow clinical and translational research.

EHRs are mainly written mainly in textual format. They lack structure or have a structure depending on the hospital, service or even the physician generated them. They contain abbreviations and metrics and are written in the language of the country. Due to unstructured nature of information locked in EHRs, detection and extraction of useful information is still a challenge and consequences in difficulty of performing Q&A process [17].

To encode, structure and extract information from EHRs, an NLP system for which the Named Entity Recognition (NER) process is its paramount task, is required. Rule-based approaches for performing NER process through means of knowledge engineering are very accurate since they are based on physician's knowledge and experience [18].

The NER process intrinsically relies on ontologies, taxonomies and controlled vocabularies. Examples of such vocabularies are Systematized Nomenclature of Medicine (SNOMED) [19] and Unified Medical Language System (UMLS) [20]. The UMLS integrates and distributes key terminology, classification and coding standards. Even though that the translations of these vocabularies to different languages are available, they do not always provide the entire terminologies that are used in very specific domains (e.g., lung cancer). In addition, several medical metrics are not covered or fully provided by them. Furthermore, symbols such as "+" and "−", which are commonly being used with medical metrics (e.g., EGFR+) to determine their positivity or negativity, are not supported by them. Also, it is a common practice by physicians to use symbols such as ".", "_", "-", etc. for writing metrics (e.g., cancer stage I-A1). Such metrics are not supported by these ontologies as well.

Although, several NLP systems have been developed to extract information from clinical text such as Apache cTAKES [21], MEDLEE [22], MedTAS/P [23], HITEx [24], MetaMap [25], to name a few. However, despite of the fact that Spanish language has occupied the second position in the world ranking of number of speakers with more than 572 million speakers [26], these systems are mainly being used for English. One of the NLP systems that has been developed to perform IE on Spanish clinical text, is C-liKES (Clinical Knowledge Extraction System) [27]. C-liKES is a framework that has been developed on top of Apache UIMA, which has been based on a legacy system, named H2A [28].

To the best of our knowledge, there is no open NLP pipeline from which we can extract information related to lung cancer mutation status, tumor stage and PS, written in Spanish clinical narratives. Thus, the main contribution of this paper is to discuss the design, development and the implementation of annotators, capable of detecting clinical information from EHRs, using UIMA framework. Furthermore, we present the annotation results, extracted by means of running these annotators. The rest of paper is organized as follows: in Sect. 2, concept annotation for mutation status, stage and PS in lung cancer domain along with annotation output generation is presented; and in Sect. 3 the achievements gained so far are explained and the outlook of the future developments is provided.

2   Solution

We have developed a set of semantic rule-based NLP modules, named Annotators, using Apache UIMA framework. These annotators were developed to identify NEs (Named Entities) from clinical narratives. They contain regular expressions, using which, they can search for specific patterns through the clinical text.

The Pseudocode of algorithms implemented by these annotators is provided in Listing. 1.

```
Start;
  Set search-pattern = regex();
  Set input = EHR plain text;
  For sentences in input {
    For tokens in sentence {
      If search-pattern is matched to a token {
        Set annotation = define the token with its corre-
sponding semantical meaning or category;
        Set output = add annotation to token indexes;
      }
      Move to the next token;
    }
    Move to the next sentence;
  }
End;
```

Listing. 1. Pseudocode of rule-based annotators algorithm

The algorithm defines the search pattern using regular expressions and accept EHR plain text as input. Then, for each individual sentence in the text, the algorithm checks if the search pattern can be matched with the tokens. Once, a matched token is found, the algorithm assigns the semantical meaning corresponding to the token and produces the output by adding annotation to the token indexes.

To process a clinical narrative using these developed annotators, we have implemented them under a single pipeline. Once, the pipeline is executed, the output of annotations will be generated.

The outcomes of annotation processes are formatted as a set of XML Metadata Interchange (XMI) files and are also inserted into a relational database from which Q&A process can be followed. The details of lung cancer developed annotators and the output generation process are provided below.

2.1   Mutation Status

Physicians makes use of EGFR, ALK and ROS1 metrics for mentioning the tumor mutation status in clinical narratives. However, in case of EGFR, they can provide more detailed information about the mutation related to the exon (18–21), type of exon (deletion or insertion) and the mutation point (G719X, T790 M, L858R, L861Q).

For determining the positivity or negativity of mutation metrics, physicians do not follow any standard systems. For example, in case of EGFR positive, they can write: *"EGFR: positive"*, *"EGFR+"*, *"has detected with mutation in EGFR"*, *"presence of mutation in EGFR"*, *"with insertion mutation in Exon 19"*, *"EGFR mutated"*, etc. Therefore, the need of annotators for detecting tumor mutation status from clinical text, comes to the picture. For this purpose, three annotators, named EGFR Annotator, ALK Annotator and ROS1 Annotator were developed using UIMA framework.

The EGFR annotator is capable of detecting the mutation status, exon, type of exon and the mutation point from the clinical text by incorporating four internal annotators that were developed for this purpose. Whereas the ALK and ROS1 annotators can only find the concepts that are related to the mutation status.

For example, in the following clinical text: *"EGFR + (del exon 19), no se detecta traslocación de ALK y ROS1 no traslocado."* while EGFR mutation status is positive in exon 19, no translocation is detected for ALK and ROS1. Once, we process this clinical text using the developed annotators for tumor mutation status on UIMA CAS Visual Debugger (CVD), this information is extracted from the text (Fig. 1).

The CVD output representation is largely divided into two sections:

- Analysis Results: is composed of two subsections: (1) upper division: contains CAS Index Repository. *AnnotationIndex* represents the list of annotators executed for processing the text, which shows *DocumentAnnotation* objects, with one specific object for each specific annotation (ALK, EGFR, Exon, …). To see the annotation results of a specific annotator, the user should click on the designated one in here; and (2) lower division: includes AnnotationIndex. When the user clicks on the index of an annotation (in the Fig. 1, -ALK-) the information such as begin, end and semantic categories of the found NE will be represented.

- Text: accepts the clinical text as input from the user, which is in here *"EGFR + (del exon 19), no se detecta traslocación de ALK y ROS1 no traslocado."*. The input text will be highlighted corresponding to the begin and end of annotation provided in the AnnotationIndex of the lower division sub-section of the Analysis Results section. The highlighted text is *"no se detecta traslocación de ALK"*.

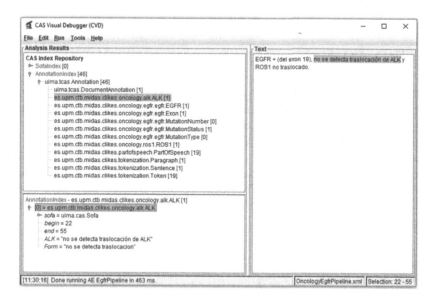

Fig. 1. EGFR, ALK and ROS1 annotation results on CVD

2.2   Stage

Stage grouping and TNM are the two main standard cancer staging systems, introduced by AJCC manual [29]. The lung cancer stage classification, which is provided by the International Association for the Study of Lung Cancer (IASLC), is based on advanced statistical analysis of international database with more than 100,000 patients. This analysis specifically addressed the stage groups, T, N and M components (Fig. 2). It is notable that for TNM system, the three attributes, i.e. T, N and M can be modulated by prefixes (e.g., pT1aN0M0). Such prefixes are: "c" (clinical), "p" (pathologic), "yc" or "yp" (post therapy), "r" (retreatment) and "a" (autopsy). Physicians normally use symbols such as ".", "_", "-", "()", etc. combined with TNM and stages metrics in the clinical text. For example, in case of stage IA1, they can write: *I-A1, I.A1, I_A_1, I(A1)*, etc.

T/M	Subcategory	N0	N1	N2	N3
T1	T1a	IA1	IIB	IIIA	IIIB
	T1b	IA2	IIB	IIIA	IIIB
	T1c	IA3	IIB	IIIA	IIIB
T2	T2a	IB	IIB	IIIA	IIIB
	T2b	IIA	IIB	IIIA	IIIB
T3	T3	IIB	IIIA	IIIB	IIIC
T4	T4	IIIA	IIIA	IIIB	IIIC
M1	M1a	IVA	IVA	IVA	IVA
	M1b	IVA	IVA	IVA	IVA
	M1c	IVB	IVB	IVB	IVB

Fig. 2. AJCC 8th edition - lung cancer stage grouping and TNM system [29]

Using 8th edition of this manuals, we have developed two pattern-based extraction annotators, named Stage Annotator and TNM Annotator, for finding the cancer stages and the TNMs, appeared in the clinical narratives, respectively.

For example, the clinical text "*Adenocarcinoma de pulmón, pT1aN0M0 (micronódulos pulmonares bilaterales, linfangitis carcinomatosa, derrame pleural), estadio I_A1.*", explains that the patient is having lung cancer (Adenocarcinoma de pulmón) with pT1aN0M0 value for TNM and stage IA1. By processing this text using the Stage and TNM annotator, the information related to these two metrics has been annotated (Fig. 3).

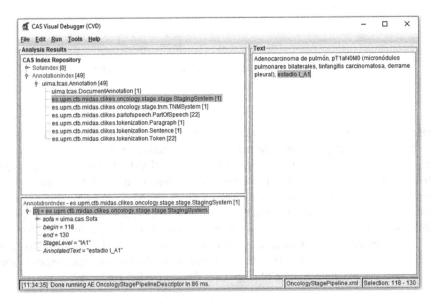

Fig. 3. Stage and TNM annotation results on CVD

2.3   PS

PS scale is mentioned using ECOG and Karnofsky measures in the clinical narratives. The ECOG measure ranges from 0 to 5, where 0 is the most ideal case for carrying on all pre-disease performance without any restrictions. On the other hand, the Karnofsky measure ranges from 0% to 100%, where 100% is the most ideal case.

ECOG and Karnfosky scales can appear with symbols such as ".", "_", "-", "()", etc. in clinical text. For example, ECOG 0 can be written as: *"ECOG_PS: 0"*, *"ECOG-0"*, *"ECOG is measured with 0"*, *"ECOG (0)"*, etc.

Hence, to annotate concepts related to ECOG and Karnofsky scales form clinical narratives, two annotators, named ECOG Annotator and Karnofsky Annotator were developed, respectively.

For example, *"ECOG-PS 0. Regular estado general. Disnea de reposo/mínimos esfuerzos. Karnofsky: 100%."*, indicates that the patient PS is measured with ECOG: 0 and Karnofsky: 100%. The results of annotation processes implemented by the ECOG and Karnofsky annotators on this clinical text, are shown in Fig. 4.

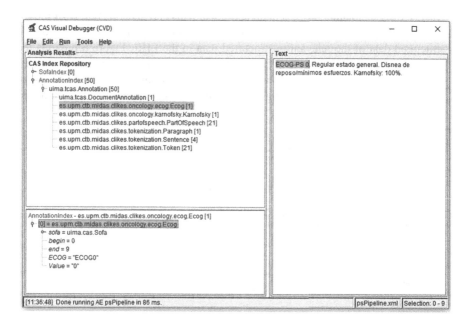

Fig. 4. ECOG and Karnofsky annotation results on CVD

2.4   Output Generation

For generating outcomes, an execution flow (Fig. 5) is followed by using a main process, called "Processing Engine". This process accepts plain text files as input. Example of plain text documents are EHRs, clinical notes, radiology reports, and any kind of medical textual document generated.

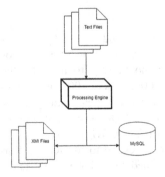

Fig. 5. Processing engine architecture

The input is passed to the Processing Engine, which contains the pipeline of developed annotators. When the Processing Engine is executed, two resources are generated as output:

- XMI: UIMA annotators process plain text documents and generate one XMI file for each of them. These files encompass all the existing annotations i.e. they contain structured data of the relevant unstructured data in the EHR. Figure 6 presents the results of annotations, stored in an XMI file, using UIMA Annotation Viewer. The UIMA annotation viewer is divided into three sections: (1) upper left division: contains plain clinical text. Highlighted tokens correspond to the annotated

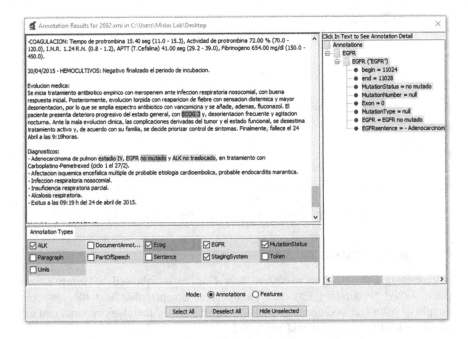

Fig. 6. XMI annotation results using UIMA annotation viewer

concepts, which are in here *"ECOG 3"*, *"estadio IV"*, *"EGFR"*, *"no mutatdo"* and *"ALK no traslocado"*; (2) Annotation Type: provides the list of annotation tags from which the user can select the annotation results to be highlighted in the upper left division section. In here such tags are ALK, ECOG, EGFR, and etc.; and (3) Click In Text to See Annotation detail: by clicking on the highlighted tokens in the upper left division, the user can see the details of annotated concepts in this section. Such details are begin, end and semantic categories of the concept.

- Structured relational database: A MySQL database, which contains the information of the annotations. The database allows to perform analysis on the structured data with more flexibility than XMI files.

3   Conclusion and Future Work

The vast amount of clinical data generated and the adaption of IT in health care industry, have motivated the development of NLP systems in clinical domain. For an NLP system to achieve a broad use, it must be capable of covering comprehensive clinical information and demonstrating effectiveness for a practical application. Thus, in this paper, we have described the development of specific case annotators for lung cancer domain, using UMIA framework. These annotators can detect information about tumor mutation status, stage of cancer and the PS from clinical text. Although, these annotators have been developed general enough so that they can be used in other oncological domains but for them to be usable in other languages, the translation of the annotator's pattern to the language is required.

This work is an on-going research, which needs further validations and developments. Such validations will go into assessment of annotation accuracy for already developed annotators whereas the developments will involve the semantic enrichment process for annotated medical concepts related to the lung cancer domain. Although, the recognition of medical concepts at NE level is one of the fundamental tasks of NLP but the judgment of clinical data cannot be understood solely at NE level. For example, clinicians can mention EGFR metric in the text for two reasons: (1) requesting for an EGFR test or (2) diagnoses of the cancer mutation status. To extract the patient's diagnosed mutation tumor status from clinical text, we need to have more semantic than NE level. Therefore, semantic enrichment process needs a great attention.

References

1. Cancer, World Health Organization. http://www.who.int/news-room/fact-sheets/detail/cancer. Accessed 12 July 2018
2. 1 in 4 deaths caused by cancer in the EU28. http://ec.europa.eu/eurostat/web/products-press-releases/-/3-25112014-BP. Accessed 21 June 2018
3. Luengo-Fernandez, R., Leal, J., Gray, A., Sullivan, R.: Economic burden of cancer across the European Union: a population-based cost analysis. Lancet Oncol. **14**(12), 1165–1174 (2013)

4. Shlomi, D., et al.: Non-invasive early detection of malignant pulmonary nodules by FISH-based sputum test. Cancer Genet. **226–227**, 1–10 (2018)
5. Zaman, A., Bivona, T.G.: Emerging application of genomics-guided therapeutics in personalized lung cancer treatment. Ann. Transl. Med. **6**(9), 160 (2018)
6. Molecular profiling of lung cancer - my cancer genome. https://www.mycancergenome.org/content/disease/lung-cancer/. Accessed 21 June 2018
7. NCI Dictionary of Cancer Terms, National Cancer Institute. https://www.cancer.gov/publications/dictionaries/cancer-terms. Accessed 21 June 2018
8. Ahmadzada, T., Kao, S., Reid, G., Boyer, M., Mahar, A., Cooper, W.: An update on predictive biomarkers for treatment selection in non-small cell lung cancer. J. Clin. Med. **7**(6), 153 (2018)
9. Oser, M.G., Niederst, M.J., Sequist, L.V., Engelman, J.A.: Transformation from non-small-cell lung cancer to small-cell lung cancer: molecular drivers and cells of origin. Lancet Oncol. **16**(4), e165–e172 (2015)
10. Iwahara, T., et al.: Molecular characterization of ALK, a receptor tyrosine kinase expressed specifically in the nervous system. Oncogene **14**(4), 439–449 (1997)
11. Rimkunas, V.M., et al.: Analysis of receptor tyrosine kinase ROS1-positive tumors in non-small cell lung cancer: identification of a FIG-ROS1 fusion. Clin. Cancer Res. **18**(16), 4449–4457 (2012)
12. AJCC - Implementation of AJCC 8th Edition Cancer Staging System. https://cancerstaging.org/About/news/Pages/Implementation-of-AJCC-8th-Edition-Cancer-Staging-System.aspx. Accessed 14 Mar 2018
13. Detterbeck, F.C., Boffa, D.J., Kim, A.W., Tanoue, L.T.: The eighth edition lung cancer stage classification. Chest **151**(1), 193–203 (2017)
14. Mak, K.S., et al.: Defining a standard set of patient-centred outcomes for lung cancer. Eur. Respir. J. **48**(3), 852–860 (2016)
15. Performance scales: Karnofsky & ECOG scores practice tools| OncologyPRO. https://oncologypro.esmo.org/Oncology-in-Practice/Practice-Tools/Performance-Scales. Accessed 12 July 2018
16. Oken, M.M., et al.: Toxicity and response criteria of the Eastern Cooperative Oncology Group. Am. J. Clin. Oncol. **5**(6), 649–655 (1982)
17. Hanauer, D.A., Mei, Q., Law, J., Khanna, R., Zheng, K.: Supporting information retrieval from electronic health records: a report of University of Michigan's nine-year experience in developing and using the Electronic Medical Record Search Engine (EMERSE). J. Biomed. Inform. **55**, 290–300 (2015)
18. Wang, Y., et al.: Clinical information extraction applications: a literature review. J. Biomed. Inform. **77**, 34–49 (2018)
19. SNOMED International. https://www.snomed.org/. Accessed 13 July 2018
20. Unified Medical Language System (UMLS). https://www.nlm.nih.gov/research/umls/. Accessed 04 May 2018
21. Savova, G.K., et al.: Mayo clinical Text Analysis and Knowledge Extraction System (cTAKES): architecture, component evaluation and applications. J. Am. Med. Inform. Assoc. **17**(5), 507–513 (2010)
22. Friedman, C., Hripcsak, G., DuMouchel, W., Johnson, S.B., Clayton, P.D.: Natural language processing in an operational clinical information system. Nat. Lang. Eng. **1**(1), 83–108 (1995)
23. Coden, A., et al.: Automatically extracting cancer disease characteristics from pathology reports into a Disease Knowledge Representation Model. J. Biomed. Inform. **42**(5), 937–949 (2009)

24. Zeng, Q.T., Goryachev, S., Weiss, S., Sordo, M., Murphy, S.N., Lazarus, R.: Extracting principal diagnosis, co-morbidity and smoking status for asthma research: evaluation of a natural language processing system. BMC Med. Inform. Decis. Mak. **6**, 30 (2006)
25. Aronson, A.R.: Effective mapping of biomedical text to the UMLS Metathesaurus: the MetaMap program. In: Proceedings of the AMIA Symposium, pp. 17–21 (2001)
26. de la Concha, V.G., et al.: EL ESPAÑOL: UNA LENGUA VIVA
27. Menasalvas Ruiz, E., et al.: Profiling lung cancer patients using electronic health records. J. Med. Syst. **42**(7), 126 (2018)
28. Menasalvas, E., Rodriguez-Gonzalez, A., Costumero, R., Ambit, H., Gonzalo, C.: Clinical narrative analytics challenges. In: Flores, V., et al. (eds.) IJCRS 2016. LNCS (LNAI), vol. 9920, pp. 23–32. Springer, Cham (2016). https://doi.org/10.1007/978-3-319-47160-0_2
29. Detterbeck, F.C.: The eighth edition TNM stage classification for lung cancer: what does it mean on main street? J. Thorac. Cardiovasc. Surg. **155**(1), 356–359 (2018)

Linked Data Based Multi-omics Integration and Visualization for Cancer Decision Networks

Alokkumar Jha[✉], Yasar Khan, Qaiser Mehmood,
Dietrich Rebholz-Schuhmann, and Ratnesh Sahay

Insight Centre for Data Analytics, National University of Ireland Galway,
Galway, Ireland
{alokkumar.jha,yasar.khan,qaiser.mehmood,
dietrich.rebholz-schuhmann,ratnesh.sahay}@insight-centre.org

Abstract. Visualization of Gene Expression (GE) is a challenging task
since the number of genes and their associations are difficult to predict in
various set of biological studies. GE could be used to understand tissue-
gene-protein relationships. Currently, Heatmaps is the standard visualization technique to depict GE data. However, Heatmaps only covers
the cluster of highly dense regions. It does not provide the Interaction,
Functional Annotation and pooled understanding from higher to lower
expression. In the present paper, we propose a graph-based technique -
based on color encoding from higher to lower expression map, along with
the functional annotation. This visualization technique is highly interactive (HeatMaps are mainly static maps). The visualization system here
explains the association between overlapping genes with and without tissues types. Traditional visualization techniques (viz-Heatmaps) generally
explain each of the association in distinct maps. For example, overlapping genes and their interactions, based on co-expression and expression
cut off are three distinct Heatmaps. We demonstrate the usability using
ortholog study of GE and visualize GE using GExpressionMap. We further compare and benchmark our approach with the existing visualization techniques. It also reduces the task to cluster the expressed gene networks further to understand the over/under expression. Further, it provides the interaction based on co-expression network which itself creates
co-expression clusters. GExpressionMap provides a unique graph-based
visualization for GE data with their functional annotation and associated
interaction among the DEGs (Differentially Expressed Genes).

1 Introduction

RNA seq and microarray data generate DEGs with their associated expression value as RPKM counts. GE is primarily responsible for gene silencing and
enhancing control by transcription initiation [5]. These genes need to be investigated to dissect the role of GE in cancer, through the networks based on their
involvement. Understanding of genes could be achieved by the integration of GE

© Springer Nature Switzerland AG 2019
S. Auer and M.-E. Vidal (Eds.): DILS 2018, LNBI 11371, pp. 164–181, 2019.
https://doi.org/10.1007/978-3-030-06016-9_16

and network data to prioritize disease-associated genes [24]. GE data is crucial to visualize, since the overall data is pattern driven where over/under expression drives the function of a gene. These functions could be understood provided associated functional annotation and GO terms could be displayed along with visualization. Secondly, most of the methods use Heatmaps to visualize GE, as a static representation where each information, such as Gene-Gene association, regulation and co-expression requires distinct visualization. To obtain interference for concluding the overall process of a gene, manual interpretation of the gene using distinct visualization becomes essential. Further, this retrieved knowledge required to be annotated for understanding the mechanism and associated cell cycle processes. We have demonstrated a graph-based method to visualize GE data. Graph-based methods have an added advantage over Heatmaps based visualization regarding GE, such as the basics of Heatmaps visualization is to define the similarity among the group of genes to build a co-expression network [15]. This tool also kept the basic requirement intact by keeping the color annotation based expression visualization as in the case of Heatmaps. Here reduction from darker to lighter color representation explains the higher to lower expression of the genes. Along with this it also generates intermediate interaction graph among transcripts or genes. The key advantage of such a mechanism is to understand the gene association, cluster with a maximum number of disease association and identify the group of critical transcripts associated with the disease or normal condition. One potential advantage could be in knockdown studies where genes group based on expression level could be used for experimental validation to understand the oncogenic properties of the gene. Another advantage could be understood by the use-case presented in this paper where we have demonstrated the relationship between the expression data of human and mouse.

2   Background

Visualization of GE is key due to its functional relevance in cancer research and other diseases. However, the development of visualization and providing a scientific source such as a mathematical model, functional annotation and associated biological process will make the task of data analytics more structured. The functional annotation will also help to map down other associated biological events like gene fusion, CNV, Methylation to develop scientifically. Since the GExpressionMap approach is mathematical model driven, integration of these concepts to build data-driven discovery will be less cumbersome. The current approaches in GE are Principle Component Analysis (PCA) plot and Box plot in general. As demonstrated in Fig. 1 which explains the pros and cons with three existing methods for GE visualization. The first method is the PCA method where principal component analysis has to be performed on the list of genes or transcripts. The outcome is usually being presented using M-A plot [27]. Such plots are primarily useful when working on a limited set of genes, as this approach radically decreases the density of visualization. Further, this visualization is chunky and adding a reference line could be challenging. When working on the RNA seq data where each experiment returns approximately 50000 transcripts and in this case reduction of dimensionality becomes essential. Sometimes due to biased analysis, there are many more "variables" than "observations". Along with this, these types of diagrams are generated either by using 'R' or 'MATLAB' which works great with smaller data sets. However, with high throughput data, it creates several issues. Further, the key to any biological outcome its functional annotation and understanding the pattern of the outcome. It is tough to accommodate functional annotation with PCA plot since data points are not so well distinguished. PCA plot is mostly static and supports limited clustering. However, it does not support the functional clustering of genes and is tough to identify sharp data points. Sometimes it is tough to distinguish two distinct clusters if they have a higher amount of overlaps. As demonstrated in Fig. 1, another method associated with GE visualization is Heatmaps based visualization. Heatmaps are called as intensity plot or matrix plot, which includes dendrogram and extended Heatmaps as well [1]. As Fig. 1 explains, it is a tabular view of a collection of data points, where rows represent genes, columns represent array experiments, and cells represent the measured intensity value or ratio. In GE visualization, Heatmaps provide multi-hue color maps for up- and down-regulation in combination with clustering to place similar profiles next to each other. Other extended versions of these Heatmaps are dendrogram, hierarchical clustering of genes or experiments, often combined with Heatmaps to provide more information about the cluster structures. The critical issue with Heatmaps for GE are, though it provides cluster structure, it is still far from the functional grouping of these clusters due to lack of integrated annotation and GO terms. It also has issues, such as it only supports qualitative interpretation possible due to color coding. It grows vertically with every additional profile and grows horizontally with every additional sample. These problems make the knowledge mining difficult for large-scale data sets, such as RNA-seq GE data. Crucial third method to visualize GE data, as shown in Fig. 1, is a one-dimensional box plot approach. Essentially this method is used for a summary of distribution, comparison of several distributions and to see the result of normalization in differentially expressed genes. This visualization is vital to understand the sample-wise or gene-wise distribution. However, due to its 1-D nature, it does not support the multiple data types represented on a single plot. For example, for a single queried gene box can plot the cut-off for the expression. However, it will not provide the entities associated with it, such as overexpressed, underexpressed and not expressed genes. This type of plots are mainly static and as explained in Fig. 1 any overlay-ed information, in this case, mutated genes for EGFR. The data points are rich even for the normalized data that it becomes tough to identify the participating entities with each gene.

3   Related Work

There are rich set of tools and web applications to visualize GE data and its biological and functional associations. The most related tools for GE visual-

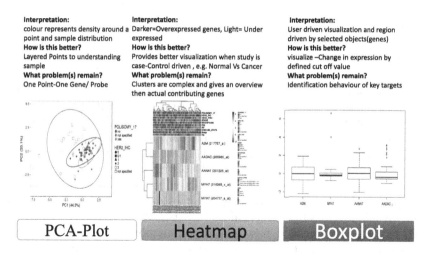

Interpretation:
colour represents density around a point and sample distribution
How is this better?
Layered Points to understanding sample
What problem(s) remain?
One Point-One Gene/ Probe

Interpretation:
Darker=Overexpressed genes, Light= Under expressed
How is this better?
Provides better visualization when study is case-Control driven , e.g. Normal Vs Cancer
What problem(s) remain?
Clusters are complex and gives an overview then actual contributing genes

Interpretation:
User driven visualization and region driven by selected objects(genes)
How is this better?
visualize –Change in expression by defined cut off value
What problem(s) remain?
Identification behaviour of key targets

PCA-Plot   Heatmap   Boxplot

Fig. 1. Motivational scenario to develop GExpressionMap for breast cancer data from E-GEOD-29431 (Color figure online)

ization are M-A plot, Heatmaps, Scree Plot, Box Plot, Scatter Plot, Wiggle Plot, Profile Plot (also known as Parallel Coordinate Plot), VA Enhanced Profile Plots and Dendrogram. In general, Mayday [6], ClustVis [17], GENE-E[1], MISO [13] are some of the most commonly used tools which covers these plots for GE. BiCluster [8] represents GE data by the hybrid approach of Heatmaps and Parallel Coordinate Plots. These plots are interactive, and GE annotations have been formalized with proper color annotations. However, this tool works on Heatmaps, and with massive data points, clusters generated by this tool can only help to infer the functionally enriched region. However, the role of each participating member and their co-expressed expressions cannot be determined. The unavailability of functional annotation and GO terms make it difficult to understand the biological processed involved with each cluster thus the pattern of the expression. INVEX [25] is again a Heatmaps based tool which deals with GE and metabolomics datasets generated from clinical samples and associated metadata, such as phenotype, donor, gender, etc. It is a web-based tool where data size has certain limitations. However, it has built-in support for gene/metabolite annotation along with Heatmaps builder. The Heatmaps builder primarily works on 'R' APIs. Though these tools have great potential due to inbuilt functional annotation, lack of clustering, interactive selection of gene entities and support for large-scale datasets provides further room for improvement. GeneXPress [20] has been developed to improve the functional annotation to reduce the task of post-processing after the obtained list of DEGs. It also contains an integrated clustering algorithm to explore the various binding sites from DEGs. Multi-view representation, which includes graph based interaction map for selected genes and Heatmaps based visualization with functional annotation, makes it a most

---

[1] http://www.broadinstitute.org/cancer/software/GENE-E/index.html.

relevant tool for GE based biological discovery with an integrated motif discovery environment. However, a single source of functional annotation raises the requirement for linked functional annotation. Again the graph visualization is limited to selected genes wherein Heatmaps identification of exact data point is a cumbersome task. GEPAT [23] is also a gene expression visualization tool developed over the Heatmaps and focused on visualization of pathway associated gene expression data. Integrated GO terms enrichment environment makes the tool unique regarding understanding the mechanism of differentially expressed genes. However, the functional annotation is mostly performed manually to have the exact mapping of each transcript involved with a certain loop of biological processed. Integrated cluster maker provides substantial support to the idea of GExpressionMap. ArrayCluster [28] is a tool developed from GE datasets, keeping in mind to resolve analytical and statistical problems associated with data. The ideal co-expression based clustering method and functional annotation of each cluster make it unique and provide a ground for GExpressionMap to include co-expression based clustering of DEGs. However, it has limited support to microarray data and makes it difficult to apply on larger gene sets generated from RNA Seq. Also, it is a Heatmaps based plotting, which makes data point selection difficult. J-Express[2] is again GE data analysis tool which contains almost every type of plot. The inclusion of various plots makes visual analytics from this tool robust. However, most of the plots generated from J-Express are static, thus lacks the key feature to understand the in-depth analysis of each tool. Integrated Gene set enrichment analysis (GSEA), Chromosome (DNA sequence) mapping and analysis, Gaussian kernels and Cross-data class prediction are some of the critical features, which makes this tool unique among others. Tang et al. presented [22] is one of the most earlier interactive visualization tool developed on the concept of ROI (Region of Interest) accommodated using scattered map. Visualization is widely supported with mathematical modeling of GE data for limited data points. This tool provides a strong foundation for GExpressionMap where we have mathematically modeled gene expression for dense and large data sets of transcripts.

4   Mathematical Model of GE and Visualization

It is essential to know the spectrum of visualization and behavior of visualizing events. Mathematical modeling of both provides a stable visualization system. Few attempts have been made earlier to model gene expression. Here we have modeled GE based on our requirement where we have identified the up-regulated, down-regulated and not expressed states for the genes and we have used it to identify meaningful data points in the cluster have an accurate co-expression network generated from GE data. GE data are a linear transcription model follows a system of differential equations [3]. The basic understanding of the terms are as follows; **Gene Expression:** Combination of genes code for proteins that are

---

essential for the development and functioning of a cell or organism. **Transcript-based co-expression network:** Set of genes, proteins, small molecules, and their mutual regulatory interactions.

The modeling could be understood by Fig. 2. As per the figure, the system could be realized as

$$\frac{\partial r}{\partial t} = f(p) - Vr, \frac{\partial p}{\partial t} = Lr - Up \tag{1}$$

where $V, U = relative degradation, L = Translation, r = concentration of gene, p = concentration of protein.

To define the over, under and no expression, and stability of cluster based on interaction network, let's assume that, at given time point t, if the concentration of mRNA is $x1$ and concentration of protein is $p = x2$, then this can be generalized as a continuous function.

$$x_i(t) \in R_{\geq 0} \tag{2}$$

$$\dot{x}_i(t) = f_i(x), \ 1 \leq i \leq n$$
$$Say\ x_1 = mRNA\ concentration, \tag{3}$$
$$p = x_2 = protein\ concentration$$

$$\dot{x}_1 = \kappa_1 f(x_2) - \gamma_1 x_1, \ \dot{x}_2 = \kappa_2 x_1 - \gamma_2 x_2$$
$$\kappa_1, \kappa_2 > 0\ production\ rate, \ \gamma_1, \gamma_2 > 0\ degradation\ rate \tag{4}$$

$$f(x_2) = f(p) = \frac{\theta^n}{\theta^n + x_2^n}$$

$$f(p) = \frac{\theta^n}{\theta^n + p^n} \tag{5}$$

if $\theta > 0$ explains genes are under expressed

else genes are over expressed

Assume $\dot{x} = 0$

$$\dot{x}_1 = 0 : x_1 = \frac{\kappa_1}{\gamma_1} f(x_2) = \frac{\kappa_1}{\gamma_1} f(p)$$

same as

$$\dot{x}_2 = 0 : x_1 = \frac{\gamma_2}{\kappa_2} x_2 \tag{6}$$

$$x_1 = \frac{\gamma_2}{\kappa_2} p$$

for x_1 and $x_2 > 0$, genes will not show expression

Another key extension of this model will be to understand the interaction model from these under/over/non expressed genes. Lets assume that these interaction networks are continuous function and cluster building follows *rate law*, then Eq. 1 can be generalized as;

$$x_i = f_i(x), where\ 1 \leq i \leq n$$
$$where f_i(x) = rate\ law\ for\ each\ interaction \tag{7}$$

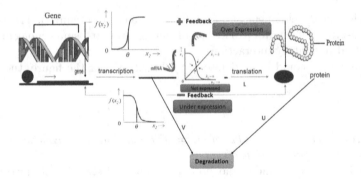

Fig. 2. Mathematical model to understand GE and gene interaction based clustering.

If translation happens with this gene then each cluster will follow a model to take part in post translational modification (PTM)s, that can be understood by,

$$p_i = f_i(p), where\ 1 \leq i \leq n \tag{8}$$

$$\frac{\partial r}{\partial t} = f_i(p) - Vr$$
$$Equation\ for\ rate\ of$$
$$change\ in\ interaction\ for\ mRNA \tag{9}$$

$$\frac{\partial [f_i(p)]}{\partial t} = Lr - Uf_i(p)$$
$$\int Lr.dt = f_i(p) - \int Uf_i(p).dt \tag{10}$$

This equation explains the relevance of GE based clustering and the effect of rate of change in expression. All the data points within this range of equation will have an easy to manageable knowledge mining. This will help to define the boundary and interpretation module from visualization.

5   Cancer Decision Networks: Integration, Model and Query Processing

Visualization is working as a presentation model with a structured and distributed model underneath for processing, filtering and querying the data. In cancer genomics, if one gene regulated by more than one events, such as gene expression, CNV, and methylation, it is unlikely that retrieved regulation occurred by chance. To realize the aspect of the multi-genomic event-based model, we have constructed a knowledge graph called "**Decision Networks (DN).**" The DN works on two-layer integration, where at first layer, we identify the linking parameters, such as Gene Symbol, CG IDs and Chr: Start-End. The detailed linked scenario is shown in Fig. 2 of [11]. However, at this level integration behaved more like an enriched dataset. Instead of building a single integrated graph, we built a virtually integrated Knowledge graph for DNs. We achieved this by federated SPARQL query as mentioned in Listing 7 of our earlier work [11]. The second layer of integration is essential regarding defining the rules to extract the biological insights from multi-omics integrated DNs. Some of the conventional rules of filtering genes without significance to make visualization clinically actionable are as follows.

(i) Gene Expression and Methylation are reciprocal to each other. Which means if the gene is hyper-methylated it should be down-regulated.

(ii) A gene cannot be up- and down-regulated at the same time.

(iii) Functional annotation follows the central dogma of disease evolution where expression is captured first and then mutation, CNV, and Methylation, respectively.

(iv) Cancer is a heterogeneous disease, and any change in one genomic event is not sufficient to understand the mechanism.

(v) Beta-value in Methylation data where negative value represents Hypo- and a positive value represents Hyper- Methylation, respectively.

(vi) The CNV, the germline DNA for a given gene, can only be risk associated it falls outside the range of USCS defined gene length.

(vii) CNV for each cancer type changes based on two parameters, namely cancer are rare **frequency** and potentially confer high penetrance called as **odds ratios**.

(vii) Any pathways represented by the change in CNV, GE and Methylation will always be given a priority in studies and thus in visualization.

After filtering the data based on rules (i–viii), as mentioned above, the systems pre-process the data as shown in Fig. 3. Figure 3 shows the key instances of input data, such as *Gene_Symbol, Chr, start, end*. The Decision Network layer we perform the integration and then visualize the filtered data. The result queried, and the filtered result can also be exported for further analysis. The use case was taken from E-GEOD-29431 - Identifying breast cancer biomarkers[3]. We have used the same genes for visualization and in Fig. 1. Figure 1 shows the data types used in the study on visualization with various techniques. Whereas Fig. 8 shows the solution on same gene as Fig. 1.

6   Functional Annotation

Integrated functional annotation is one of the key advantages associated with visual mining of GE data sets. We have used a semantic web approach to link distinct data sets from COSMIC, TCGA and ICGC. In comparison with existing data linking methods, our approach has linked data sets based on the semantics within the data. For example, we have extracted CNV, GE, Mutation and DNA

---

[3] https://www.ebi.ac.uk/arrayexpress/experiments/E-GEOD-29431/samples/.

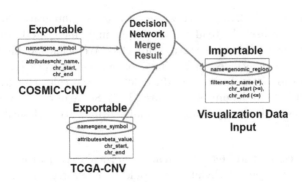

Fig. 3. Data input output model for visualization

Methylation data from TCGA, COSMIC, and ICGC and linked them to have enriched semantics, which in turn leads to having an improved coverage of the genome for each genomics profile. Each of these genomic signatures has its dedicated SPARQL endpoints. These SPARQL endpoints will be iteratively enriched with other associated similar data types to have maximum coverage of genome for each genomic profile. In the present paper, all differentially expressed genes from use case annotated using GE data from our Linked functional annotation platform [10].

Table 1. Genomics data statistics

No.	Data	Triples	Subjects	Predicates	Objects	Size (MB)
1	COSMIC GE	1184971624	48121454	18	148240680	10000
2	COSMIC GM	83275111	3620658	23	9004153	1400
3	COSMIC CNV	8633104	863332	10	921690	122
4	COSMIC Methylation	170300300	8292057	22	603135	2800
5	TCGA-OV	81188714	10974200	15	4774584	3774
6	TCGA-CESC	3763470	627652	43	481227	49557
7	TCGA-UCEC	553271744	19233824	91	68370614	84687
8	TCGA-UCS	1120873	183602	36	188970	10018
9	KEGG	50197150	6533307	141	6792319	4302
10	REACTOME	12471494	2465218	237	4218300	957
11	GOA	28058541	5950074	36	6575678	5858
12	ICGC	577 M	−−	−−	−−	39000
13	CNVD	1,552,025	194,590	9	512,307	71

Table 1 shows the overall statistics of RDFization of COSMIC, TCGA and CNVD data and external (RDF) datasets used: rows 1–4 represent the number of triples and its size for COSMIC gene expression, gene mutation, CNV and Methylation data sets, respectively. Rows 5–8 represent the number of triples, and it is size for TCGA-OV, TCGA-CESC, TCGA-UCEC and TCGA-UCS data, respectively. The RDFization statistics for CNVD data are shown in row 13. Rows 9–12 represents the statistics of external datasets (available in RDF format), namely KEGG, REACTOME, GOA, and ICGC. To query data, we have used an adapted version of SAFE [14], a federation engine to query data from multiple endpoints in a policy-driven approach which may be a key element from the user while the user is selecting his/her hypothesis from visualization and unique functional annotation module based on the distributed concept in genomics.

7   GExpressionMap

GExpressionMap has been built over a robust mathematical model of gene expression which defines that GE is linear and having a graph-based visualization for linear model provides the better visual representation of the events. In addition to visualization, we have built linked data based decision networks where we have contributed TCGA-OV, TCGA-UCS, TCGA-UCSC and TCGA-CESC (Methylation, CNV, Gene expression, and Complete Mutation) data along with COSMIC (GE, CNV, GM, and Methylation) and CNVD extending our earlier work [10–12]. These datasets will provide a platform for link identification and federation and addition to Linked Open Data.

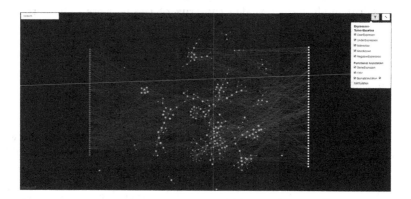

Fig. 4. The GExpressionMap main view where the left side represents the lower and right side represents the higher gene expression (Color figure online)

GExpressionMap has been divided into four modules to identify critical challenges associated with GE data sets in biology. The first mode called as **Expression mode** talk about the conditional expression and track the changes in the property of transcripts or genes based on the changes in the expression level and identify the role on non-expressed genes in various cell cycle processes. Another mode called **knockdown mode** identifies the changes in various clusters representing a group or a biological process. This mode will also help to understand the effect from a knockdown to knockout. Knockdown studies are essential to solve various biological problems, such as a natural mechanism for silencing gene expression, specific inhibition of the function of any chosen target gene to understand the role in cancer and other diseases. Tracking these changes in the graph-based on motif building or destructing and cluster changes provides a visual impact to this biological discovery [18]. Another critical challenge while dealing with the group of genes or transcripts is to understand the pattern or bias of the network/data to understand the mechanism of the experiment. GExpression-Map provides an integrated annotated genes with their functional annotation and further cluster them based on their RPKM values means their expression pattern. By this way, the experimenter will conclude that how reliable is a cleave from a cluster what functional processes they are involved in and what can be cumulative effect reported from validated and patient data sources based on the linked functional annotation and GO annotations. This dimension of work is called **Annotation, Clustering and GO processes** mode. It is always crucial to find the strongest and weakest cluster based on matrices, such as the number of overexpressed genes connected with a cluster, number of underexpressed genes connected with a cluster or participation of individual gene in a cluster. On the other hand, if critical genes, such as TP53, EGFR, BRCA, and other biomarker is associated with large no of network or clusters can drive the progression in the disease like cancer. However, the number of over and under-expressed genes with this network will explain the functioning. This is how **Interaction and co-expression** mode have revealed the crux of the network. The aerial view of the expression map is depicted in Fig. 4. Details of each mode explained in following subsections.

7.1 Expression Mode

Expression mode overlays the GE data either from microarray or RNA seq based on RPKM count from lower to the higher expression. As explained in Fig. 4 red color bar demonstrated the gene with lower expression value whereas the white expression bar explains the value with higher expression value. The list of bubbles is the genes are either highly or lowest expressed based on their expression value. The expression scale in the bottom is the log scale which explains the range of expression considered maximum to minimum as RPKM/FPKM values. As it can be observed from Fig. 6 that bottom expression line annotated as **D** is being used in such a way to have two-way side slider pointer. The major use of this approach is to identify the most significant genes since the expression value from RNA seq has a broader range mostly. The Value as mentioned in **A** explains about two types of values annotated as *OE*-Overexpressed and UE-Under-expressed. The value is constantly displayed as per the change from slider annotated on Fig. 6 as **B**. The example of this has been shown as **C** where the for value 49848 expressed of *PSMD9* having been displayed. The overall impact of this mode would be to retain the ease of expression scale as in the case of Heatmaps, however covering the broad spectrum of the gene with added functionalities.

7.2 Knockdown Mode

Knock-down studies play a key role in biological experiments to understand the overall impact of a *gene* or *mRNA*. For example in cancer networks where if we consider one GE network contains the expression interaction from normal and adjacent normal called as normal sample expression network. Another network could be the expression network obtained from cancer tissues. Now to understand the behavior is important to understand the knockdown effect of most affected genes. As GExpressionMap also provides *bottleneck genes* based on cluster binding and a number of the associated cluster with that gene, the strength of the cluster. If a single gene has different expression level in both normal and cancer network, it would be key to understand the impact of losing that gene and then understand the overall pattern of the network. Especially cancer network can get distorted after losing these bottleneck gene or highly expressed genes. A key observation such as the presence of certain genes with higher cluster binding in normal network however absence in cancer network can lead to key outcomes in cancer studies. Figure 5 provides a snippet of one such case. As explained in the figure knockdown of *PSMD9* will affect two genes from higher expression pole and two genes from a lower expression pole. Further, the cluster associated with it and having lower expression will have loss of connectivity and will cause insatiability in the network. This is a typical example of cancer progression and loss of connectivity in the cancer networks.

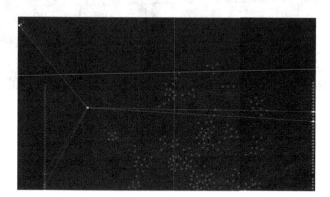

Fig. 5. A bottleneck view to understanding the effect of expression change and associations

7.3 Interaction and Co-expression Mode

Dynamic changing property from normal to cancer networks reveals common system-level properties and molecular properties of prognostic genes across cancer types [26]. However current methods to generate co-expression network are basically for microarray data since they have been defined based on probe ids. This types of the network will not be able to cope-up to identify the changes in the broader level in co-expression networks [9]. This paper builds the co-expression network based on raw RPKM/FPKM values, or it can also accommodate expression value as log2 fold change values [16]. One of the key impacts of building a co-expression network using these expression counts is to bring similar associations or cell functions together after clustering. Usually in cancer networks, one of the major issues is to identify missing links and predict the fill-ins for the missing links. Since the RPKM values are experiment specific becomes essential to track the change and loss of expression for same tissue across different experiments. Building a co-expression network by this approach will automatically define the causality of the network if changes are abrupt. If a certain transcript is not at all expressed or lost the connectivity due to some treatment in any of the control would be easy to track. Apart from this differentially expressed genes could be easily extended to differentially expressed pathways based on co-expression network. This could be one of the potential outcomes. Figure 6 provides a glimpse of a co-expression network. One of the key points in this visualization is that it highlights the high expression network and keeps the less expressed network in light color annotations. Figure 6 clearly indicates that cluster A is highly expressed than B,C,D among these co-expressed networks.

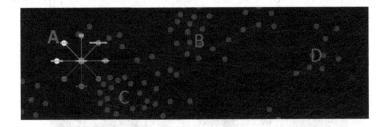

Fig. 6. GExpressionMap leveling and interacting partner association to visually mine functional annotations. (Color figure online)

7.4   Annotation, Clustering and GO Processes Mode

This mode of GExpressionMap involves the key features such as retrieval of *GO:ID* for a bottleneck gene identified based on clustering. To reduce the complexity in the visualization GExpressionMap have placed annotation based on user request. As depicted in Fig. 6 where **C** indicates the bottleneck since holding three expression cluster. Now if the user is interested in functional annotation of this gene, they need to retrieve GO biological process and as mentioned in Fig. 4 as *G* clicking on this would provide a to an interface to obtain annotations as displayed in Fig. 7. As we click on **G** of Fig. 6 it takes to the **a** of Fig. 7 and user need to enter the bottleneck gene obtained. Then interface queried a flat file for annotations [4]. This will display Go Ids and other Ids associated with the input query gene represented as **b** in Fig. 7. Once we have obtained the GO Ids we have used gene ontology search engine obtained from[4] and embedded with our system. Then we query for obtained GO Id and outcome of some can be represented as **d** and **e** in Fig. 7. This way we have contributed a web application with visualization to annotated the gene with associated expression visualization and identification of bottleneck gene or protein. Another key is to identification and understanding of clusters. One of such cluster based on our use case having been shown in Fig. 8. The details of these clusters and associated methods will be discussed in the Result section.

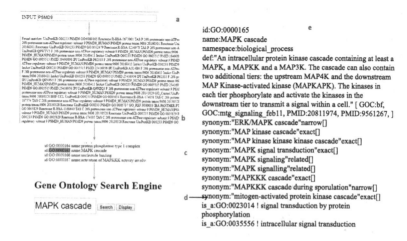

Fig. 7. The Go ontology and functional annotation for the human-mouse model use case.

8   Case Study, Results and Discussion

To demonstrate the feasibility of the proposed approach in biology, we have demonstrated a use-case from Monaco et al. [19]. This paper represents the comparative GE data between human and mouse. We have used GExpressedMap to visualize this data and draw some of the key conclusions using visual representation. Based on the steps mentioned earlier, we have developed an expression map where Fig. 8 represents one of the key clusters from this expression map. As we can observe from the diagram, human genes *A2M* have a close expression concerning mouse genes such as *Aanat, Aadac, Amap, Abat, Abca1, and Aars*. Here, the key observation is that this cluster also holds other clusters and

---

[4] https://github.com/zweiein/pyGOsite.

becomes bottleneck genes in human-mouse expression network. On the other hand, the only A2M human gene is underexpressed, and has a strong correlation with underexpressed genes in mouse (such as *Aanat, Aadac, Amap, Abat, Abca1*) as well as an overexpressed gene in mouse (such as *Aars*). One of the key outcomes of this cluster could be to identify detectable expression differences between species or individuals. The expression could logically divided into selectively neutral (or nearly neutral) differences and those underlying observable phenotypic [7]. To dig in further to identify the fact we have extracted the GO ids for each of the genes involved in the cluster. Where A2M highly associated with GO terms such as GO:0003824, GO:0004867, GO:0010951 and GO:0070062. Where, GO:0003824 is responsible for *catalytic activity* and has close correlation with GO:0003674, GO:0004867 associated with *serine-type endopeptidase inhibitor activity* and has close association with GO:0004866: *endopeptidase inhibitor activity*, whereas GO:0010951 and GO:0070062 are associated with negative regulation of endopeptidase activity and extracellular exosome respectively. To establish an association between human-mouse cluster, we have used the MGD [2] database, as the current version of GExpressionMap only supports *homospaiens*. The annotations for *Aanat, Aadac, Amap, Abat, Abca1* are a protein-coding gene which has the relation of *A2M*. These genes Aanat(cellular response to cAMP circadian rhythm, melatonin biosynthetic process, N-terminal protein amino acid acetylation), Aadac (carboxylic ester hydrolase activity, deacetylase activity, endoplasmic reticulum, endoplasmic reticulum membrane), Amap, Abat (aging, behavioral response to cocaine, catalytic activity, copulation), Abca1 (anion transmembrane transporter activity, apolipoprotein A-I binding, apolipoprotein A-I-mediated signaling pathway, apolipoprotein A-I receptor activity). Where the only highly expressed gene in mouse *Aars(alanine-tRNA ligase activity, cellular response to unfolded protein, skin development, tRNA modification)* having relation with *A2M*. Based on the biological process, this cluster represents *Membranoproliferative Glomerulonephritis, X-Linked Tangier Disease; TGD* and A2M are also involved with X-Linked Tangier Disease. In Summary, the visual identification of cluster, mapping of GE for each associated gene with the cluster, identification of expression level and functional annotation provides a key solution to how orthologs data with GExpressionMap have helped to mine the gene association to predict possible disease based on expression data. The proposed case study and results have just provided initial insight into a hidden treasure that can dig down visually using GExpression-Map. The expression extended for time series co-expression data where expression change happens on a certain time interval. For instance effect of ZIKA virus [21] where expression of top genes visualized for 12, 48 and 96 h.

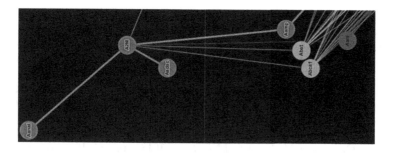

Fig. 8. Cluster representing diseasome for human-mouse

9 Conclusions

GExpressionMap is a key mechanism developed for visualization of gene expression data which is highly user-friendly, interactive, modular and visually informative. Integrated functional annotation, clustering, and co-expression network based on scientifically selected color annotations make it highly informative, usable and associative towards biological discovery based on genes expression.

Acknowledgment. This publication has emanated from research conducted with the financial support of Science Foundation Ireland (SFI) under Grant Number SFI/12/RC/2289, co-funded by the European Regional Development Fund.

References

1. Battke, F., Symons, S., Nieselt, K.: Mayday-integrative analytics for expression data. BMC Bioinform. **11**(1), 121 (2010)
2. Blake, J.A., Richardson, J.E., Bult, C.J., Kadin, J.A., Eppig, J.T.: MGD: the mouse genome database. Nucleic Acids Res. **31**(1), 193–195 (2003)
3. Chen, T., He, H.L., Church, G.M., et al.: Modeling gene expression with differential equations. In: Pacific Symposium on Biocomputing, vol. 4, p. 4 (1999)
4. Gene Ontology Consortium: Gene ontology consortium: going forward. Nucleic Acids Res. **43**(D1), D1049–D1056 (2015)
5. Delgado, M.D., León, J.: Gene expression regulation and cancer. Clin. Transl. Oncol. **8**(11), 780–787 (2006)
6. Dietzsch, J., Gehlenborg, N., Nieselt, K.: Mayday-a microarray data analysis workbench. Bioinformatics **22**(8), 1010–1012 (2006)
7. Dowell, R.D.: The similarity of gene expression between human and mouse tissues. Genome Biol. **12**(1), 101 (2011)
8. Heinrich, J., Seifert, R., Burch, M., Weiskopf, D.: BiCluster viewer: a visualization tool for analyzing gene expression data. In: Bebis, G., et al. (eds.) ISVC 2011. LNCS, vol. 6938, pp. 641–652. Springer, Heidelberg (2011). https://doi.org/10.1007/978-3-642-24028-7_59
9. Hong, S., Chen, X., Jin, L., Xiong, M.: Canonical correlation analysis for RNA-seq co-expression networks. Nucleic Acids Res. **41**(8), e95 (2013)

10. Jha, A., et al.: Linked functional annotation for differentially expressed gene (DEG) demonstrated using illumina body map 2.0. In: Malone, J., Stevens, R., Forsberg, K., Splendiani, A. (eds.) Proceedings of the 8th Semantic Web Applications and Tools for Life Sciences International Conference, CEUR Workshop Proceedings, Cambridge UK, 7–10 December 2015, vol. 1546, pp. 23–32. CEUR-WS.org (2015)

11. Jha, A., et al.: Towards precision medicine: discovering novel gynecological cancer biomarkers and pathways using linked data. J. Biomed. Semant. 8(1), 40 (2017)

12. Jha, A., Mehdi, M., Khan, Y., Mehmood, Q., Rebholz-Schuhmann, D., Sahay, R.: Drug dosage balancing using large scale multi-omics datasets. In: Wang, F., Yao, L., Luo, G. (eds.) DMAH 2016. LNCS, vol. 10186, pp. 81–100. Springer, Cham (2017). https://doi.org/10.1007/978-3-319-57741-8_6

13. Katz, Y., Wang, E.T., Airoldi, E.M., Burge, C.B.: Analysis and design of RNA sequencing experiments for identifying isoform regulation. Nat. Methods 7(12), 1009–1015 (2010)

14. Khan, Y., et al.: Safe: policy aware SPARQL query federation over RDF data cubes. In: SWAT4LS (2014)

15. Khomtchouk, B.B., Van Booven, D.J., Wahlestedt, C.: HeatmapGenerator: high performance RNAseq and microarray visualization software suite to examine differential gene expression levels using an R and C++ hybrid computational pipeline. Source Code Biol. Med. 9(1), 1 (2014)

16. Kommadath, A., et al.: Gene co-expression network analysis identifies porcine genes associated with variation in Salmonella shedding. BMC Genomics 15(1), 1 (2014)

17. Metsalu, T., Vilo, J.: ClustVis: a web tool for visualizing clustering of multivariate data using Principal Component Analysis and heatmap. Nucleic Acids Res. 43(W1), W566–W570 (2015)

18. Mocellin, S., Provenzano, M.: RNA interference: learning gene knock-down from cell physiology. J. Transl. Med. 2(1), 39 (2004)

19. Monaco, G., van Dam, S., Ribeiro, J.L.C.N., Larbi, A., de Magalhães, J.P.: A comparison of human and mouse gene co-expression networks reveals conservation and divergence at the tissue, pathway and disease levels. BMC Evol. Biol. 15(1), 259 (2015)

20. Segal, E., et al.: GeneXPress: a visualization and statistical analysis tool for gene expression and sequence data. In: Proceedings of the 11th International Conference on Intelligent Systems for Molecular Biology (ISMB), vol. 18 (2004)

21. Singh, P.K., et al.: Determination of system level alterations in host transcriptome due to Zika virus (ZIKV) Infection in retinal pigment epithelium. Sci. Rep. 8(1), 11209 (2018)

22. Tang, C., Zhang, L., Zhang, A.: Interactive visualization and analysis for gene expression data. In: Proceedings of the 35th Annual Hawaii International Conference on System Sciences, HICSS 2002, p. 9-pp. IEEE (2002)

23. Weniger, M., Engelmann, J.C., Schultz, J.: Genome Expression Pathway Analysis Tool-analysis and visualization of microarray gene expression data under genomic, proteomic and metabolic context. BMC Bioinform. 8(1), 179 (2007)

24. Wu, C., Zhu, J., Zhang, X.: Integrating gene expression and protein-protein interaction network to prioritize cancer-associated genes. BMC Bioinform. 13(1), 182 (2012)

25. Xia, J., Lyle, N.H., Mayer, M.L., Pena, O.M., Hancock, R.E.: INVEX-a web-based tool for integrative visualization of expression data. Bioinformatics 29(24), 3232–3234 (2013)

26. Yang, Y., Han, L., Yuan, Y., Li, J., Hei, N., Liang, H.: Gene co-expression network analysis reveals common system-level properties of prognostic genes across cancer types. Nat. Commun. **5**, 3231 (2014)
27. Yeung, K.Y., Ruzzo, W.L.: Principal component analysis for clustering gene expression data. Bioinformatics **17**(9), 763–774 (2001)
28. Yoshida, R., Higuchi, T., Imoto, S., Miyano, S.: ArrayCluster: an analytic tool for clustering, data visualization and module finder on gene expression profiles. Bioinformatics **22**(12), 1538–1539 (2006)

The Hannover Medical School Enterprise Clinical Research Data Warehouse: 5 Years of Experience

Svetlana Gerbel[1]([⊠])[ID], Hans Laser[1][ID], Norman Schönfeld[1][ID],
and Tobias Rassmann[2][ID]

[1] Hannover Medical School, Carl-Neuberg-Str. 1, 30625 Hannover, Germany
gerbel.svetlana@mh-hannover.de
[2] Volkswagen Financial Services AG,
Gifhorner Straße 57, 38112 Brunswick, Germany

Abstract. The reuse of routine healthcare data for research purposes is challenging not only because of the volume of the data but also because of the variety of clinical information systems. A data warehouse based approach enables researchers to use heterogeneous data sets by consolidating and aggregating data from various sources. This paper presents the Enterprise Clinical Research Data Warehouse (ECRDW) of the Hannover Medical School (MHH). ECRDW has been developed since 2011 using the Microsoft SQL Server Data Warehouse and Business Intelligence technology and operates since 2013 as an interdisciplinary platform for research relevant questions at the MHH. ECRDW incrementally integrates heterogeneous data sources and currently contains (as of 8/2018) data of more than 2,1 million distinct patients with more than 500 million single data points (diagnoses, lab results, vital signs, medical records, as well as metadata to linked data, e.g. biospecimen or images).

Keywords: Clinical Research Data Warehouse · Secondary use of clinical data
Data integration · BI · Data and process quality · Text mining
KDD · System architecture

1 Introduction

1.1 Data Warehouse and Secondary Use of Clinical Data

The secondary use of information means using the information outside the original purpose of use, e.g. using routine health care data for quality assurance or scientific purposes. The reuse of electronic health record (EHR) data for research purposes has become an important issue in the national debate [1, 2].

A typical large university hospital is characterized by a heterogeneous IT system landscape with clinical, laboratory and radiology information systems and further specialized information systems [3]. The IT system landscape of an university hospital usually includes systems for documentation and processing of data that are generated during the provision of health services (e.g. diagnostics and clinical findings) or the

© Springer Nature Switzerland AG 2019
S. Auer and M.-E. Vidal (Eds.): DILS 2018, LNBI 11371, pp. 182–194, 2019.
https://doi.org/10.1007/978-3-030-06016-9_17

administration of patient data (e.g. master data). The totality of these systems is referred to as the Hospital Information System (HIS).

The Hannover Medical School (Medizinische Hochschule Hannover, MHH) is no exception in this context. In addition to the clinical and laboratory information system (HIS and LIS), the range of IT solutions used at the MHH also includes a number of individual solutions in the field of clinical research based on a wide variety of widely used database management systems (from Oracle, Microsoft, FileMaker etc.) as well as an unmanageable number of table-based documentation systems.

A universal approach for central data integration and standardization within an organization with heterogeneous databases is to build a data warehouse system based on database component consisting of consolidated and aggregated data from different sources. Data warehouse technology allows users to run queries, compile reports, generate analysis, retrieve data in a consistent format and reduce the load on the operative systems.

Already Teasdale et al. [4] and later Bonney [5] made clear that the reuse of clinical (primary) data is facilitated by Business Intelligence on the basis of a so-called "Research Patient Data Repository" or Clinical Data Warehouse. The use of Business Intelligence creates the basis for further extraction of empirical relationships and knowledge discovery in databases (KDD), like in the field of data science.

The relief of operative data processing and application systems is another central argument for the use of data warehouse technology. This makes it possible to execute requests for clinical data on a dedicated repository rather than at the expense of the operative systems [6].

In the clinical-university sector in Germany, there is a series of established data warehouse solutions for secondary data use. They are commonly described as Clinical Data Warehouses (CDW) [2]. In the IT environment of a hospital, the response times of operative systems (e.g. clinical workplace applications) are a particularly critical factor. Accessing the operational systems with real-time queries would increase the likelihood of non-availability. In addition to these drivers, the following typical application scenarios for secondary use [2, 3, 7, 8] have formed:

– Patient screening for clinical trials based on inclusion and exclusion criteria
– Decision support through comparison of diagnosis, therapy and prognosis of similar patients
– Epidemiological evaluations by examining the development of frequencies of clinical parameters (e.g. risk factors, diagnoses, demographic data)
– Validation of data in registers and research databases and their data enrichment with the aim of quality improvement

In the review of Strasser [3] from 2010, the IT systems of 32 German university hospitals were evaluated with regard to the components of the HIS and the available data warehouse solutions for consolidating routine clinical data for secondary use of data. The results of this survey correspond to a survey conducted by the CIO-UK (Chief Information Officers - University hospitals) in 2011 in order to identify existing IT infrastructures and technology stacks in Germany that solve the challenges of data integration and data management for secondary data use with regard to Clinical Data Warehouse technology. The CIO-UK represents interests from the 35 university hospitals in Germany. In summary, it can be said that there is no universal solution to implement a data warehouse technology for secondary use in Germany. There are numerous different implementations of CDWs in the community that use open source based frameworks such as i2b2[1] (incl. tranSMART) or proprietary development solutions with popular DBMSs (for example, Microsoft SQL Server, Oracle and PostgreSQL) [8, 9]. Generally speaking, currently i2b2 and OMOP[2]-based approaches appear to be the most widely used worldwide [10].

1.2 Background

MHH is one of the most efficient medical higher education institutions in Germany. As an university hospital for supramaximal care with 1,520 beds, the MHH treats patients who are severely ill. They benefit from the fact that the medical progress developed at the university is quickly available to the patients. Every year more than 60,000 people are treated in more than 70 clinics, institutes and research facilities; in the outpatient area there are around 450,000 treatment contacts per year [11].

Centralisation of the operational systems at the MHH is ensured by the Centre for Information Management (ZIMt). The ZIMt is responsible for the provision of the operational systems and ensures maintenance, support as well as the adaptation of the IT systems to the needs of the MHH. The ZIMt operates a class TIER 3 computer centre [12] at the MHH, i.e. a primary system availability of 99.982%. In addition, the MHH departments are certified according to DIN EN ISO 9001:2015. Thus, the highest demands are placed on the processes (SOPs) and offer patients as well as employees the assurance that they can rely on compliance with defined quality standards.

The Enterprise Clinical Research Data Warehouse (ECRDW) is an interdisciplinary data integration and analysis platform for research-relevant issues that has been available enterprise-wide since July 2013 [7]. The provision and support of the ECRDW as a central service at the MHH is carried out by the Division for Educational and Scientific IT systems of the ZIMt.

2 Materials and Methods

2.1 Data Sources and Interfaces

The HIS of the MHH is operated by the ZIMt and consists of more than 50 sub-components (e.g. Electronic Medical Record System (EMS), Laboratory Information Systems (LIS) and Radiology Information Systems (RIS)), which exchange EHR data via a communication server. The ECRDW integrates the EHR data of the HIS via existing HL7 interfaces, which are provided via a communication server, as well as via separate communication paths for systems that are not or no longer connected to the communication server (legacy systems).

---

[1] https://www.i2b2.org/, https://www.i2b2.org/webclient/

[2] https://www.ohdsi.org/data-standardization/

2.2   Selection Process and Evaluation of an Appropriate Data Warehouse Development Platform

The selection of a data warehouse technology was carried out between 2010 and 2011 by a working group of ZIMt and 12 other MHH departments (a total of 24 participants from the fields of IT, clinical, biometrics and clinical trial). The multi-stage selection process was divided into: product presentations, software implementations, workshops with suppliers, (weighted) evaluation of the tools by the working group members using the developed catalogue of requirements. The decisive selection criteria (inclusion and exclusion criteria) were as follows:

- Complete solution (data integration and analysis tools)
- Independence from the product vendor (autonomous development possible)
- Powerful data integration tool (ETL)
- Active community (knowledge bases, know how, support)
- Suitable license model

A total of five software vendors were evaluated. At the end of a 10-month selection process, the working group opted for Microsoft SQL Server BI data warehouse technology based on the selection criteria. The trend described in the Magic Quadrant of Business Intelligence Platforms 2011 by the Gartner Group market research report complemented our evaluation [13].

2.3   Development Architecture and Data Warehouse Approach

The development in the ECRDW is based on a three-tier-deployment-architecture: development, test and production environment (Fig. 1). The development environment consists of a database server and a dedicated server for version control in order to develop ETL processes and store the development artifacts in a version-secured manner. Systems of the development and test environment are virtualized to more dynamically distribute and economize resources. The test environment reflects the structure of the production environment in a virtual environment and is thus divided into a database server, analysis server and reporting server. Development statuses are first delivered and tested in the test environment. After successful test runs, the development artifacts (releases) are rolled out to the production environment (rollout) and tested again. The production environment therefore consists of a database server, analysis server, and reporting server. For performance reasons, the database server in the production environment is not virtualized.

The ECRDW is based on the Microsoft SQL Server architecture. The SQL Server serves as database management system and core data warehouse repository. The additional service SQL Server Integration Services (SSIS) is used for data integration via the ETL (Extract, Transform, Load) process and for developing and providing Business Intelligence (BI) solutions SQL Server Analysis Services (SSAS). The data modelling is done in concepts in a relational data model and is based on the Inmon architecture [14]. This means that data is first merged into a consolidated layer, which forms the basis for departmental views of the data sets (known as data marts). This

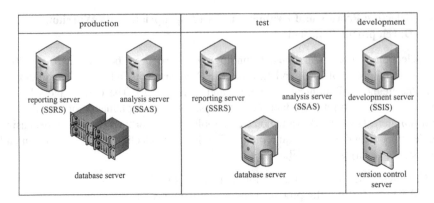

Fig. 1. The ECRDW three-tier-deployment-architecture

approach makes it possible to create a comprehensive respective global scheme for the data captured at the MHH.

Since the primary data is collected for clinical purposes and for billing, a pre-processing is necessary, e.g. to reduce the heterogeneity of the system-specific data models and to check data integrity and consistency. In addition, clinical data in a HIS is exchanged typically among the primary systems via HL7. Logically, this leads to redundant storage of information. Thus, master data transmitted via the HL7 message type ADT (Admission, Discharge, Transmission) to each primary system is duplicated. For each single fact from the clinical documentation there is a primary leading oper-ative system at the MHH. To avoid possible ambiguities or redundancies (requirement for entity matching) in data integration, the leading operative system is identified in each integration project. Information from other primary systems of the HIS that consume the primary data of another system is not integrated. If two primary systems represent a similar concept (e.g. laboratory findings), the semantic integration of both systems into the same concept of the ECRDW takes place. If duplicates of the infor-mation are produced, the primary operative system for this information is identified again. Deduplication thus takes place within the ETL process. Primary data from the various data sets of the operative systems are incrementally integrated into a central repository using data warehouse technology. Depending on the possibilities of the HIS subsystem, this process takes place daily or weekly in an incremental loading process via ETL into the ECRDW core repository. After consolidation and standardization in modelled standard concepts (e.g. historized master data), central use cases can be served by providing targeted data selections.

2.4   Methods to Ensure Data Protection

In Germany, the evaluation of primary data arising in the context of treatment is regulated by aspects of data protection at EU, federal and state level. In addition there are the legal regulations (SGB V, SGB X, infection protection, cancer early detection and cancer register, hospital laws etc.). This special feature is reflected in the possi-bilities of using these data for research purposes (secondary data use) and inevitably leads to a limitation of the use cases for such data. In secondary data analysis, the right to informational self-determination of the individual must always be protected and weighed against the right to freedom of science and research [15].

With the entry into force of the EU Basic Data Protection Regulation (GDPR) (EU 2016/679) on 25 May 2018, the processing of genetic, biometric and health data is prohibited under Article 9(1) of the GDPR, unless the person has explicitly consented to the use of the data. The current Lower Saxony Data Protection Act (§ 13 NDSG) states that a person must have given consent to the use of personal data for scientific purposes. A data protection and access concept must therefore be defined in order to comply with data protection regulations. The implementation is described in 3.4.

2.5 ECRDW Use Cases

Typical application scenarios of a clinical data warehouse [2, 7, 8] were implemented at the MHH in three central application cases.

Screening: By means of a so-called anonymous cohort identification, researchers have the possibility to define a cohort via a data warehouse on the basis of inclusion and exclusion criteria. The criteria are used to calculate quantities for e.g. patients, cases and laboratory values on the basis of the EHR data and to be able to provide a statement on the feasibility of the research question. A screening for clinical studies is possible analogously and can contribute to the reduction of the sometimes time-consuming research on EHR data [16–19].

Epidemiological Study: Similar to screening in a clinical study, data collection can also be very time-consuming when performing a retrospective data analysis (epidemiological study). Medical findings are sometimes only available as PDF documents in the central archive for patient files after completion of treatment. Through the use of a data warehouse, a wealth of information about the entirety of the patients of a hospital, the clinical pictures and the context-specific final results of the therapy (e.g. condition at discharge) can be provided, which are available in the various application systems of a HIS.

Validation and Data Enrichment: Research and registry databases often suffer from manual data entry (so-called "media discontinuity"). As a result, errors, typos, incomplete or erroneous data collection are a challenge that many such data collections have to overcome [1]. The use of data warehouse technology can make a decisive contribution to correcting errors in existing information. In addition, data from a register can be completed by adding additional data from the database of a data warehouse (e.g. risk factors from EHR data) [20].

2.6 Information Quality Scheme: Process Chain, Artefact, Relativity (PAR)

In order to improve the information quality in the analysis of large amounts of data, the criteria for mapping the information quality of Wang and Strong [21] were examined and modified for transferability to a clinical research data warehouse.

The result was a modified three-dimensional information quality scheme (Process chain, Artefact, Relativity, PAR) with a total of 26 criteria. During the development of the ECRDW, a subset of information quality criteria of the PAR scheme which are assigned to the sub-process of processing in the process chain dimension and by the definition of templates has been implemented. The aspect of reuse was the key point of the process chain dimension. These are 12 information quality criteria: standardization, source traceability, loading process traceability, processing status, reference integrity, uniform presentation, data cleansing scope, degree of historization, no-redundancy, performance and restartability [22].

3   Results

3.1   Content of the ECRDW

The MHH ECRDW has been continuously integrating data from MHH's primary systems into a relational, error-corrected and plausibility-tested data model since it went live in 2013 (see Table 1).

Table 1. Content of the ECRDW (as of July 2018)

Domains (millions)	07/2013	07/2014	07/2015	07/2016	07/2017	07/2018
Biospecimen	–	–	–	–	0,01	0,04
Demographic data	1,97	2,12	2,28	2,47	2,64	2,93
ICD diagnosis	–	6,65	7,75	8,92	9,88	11,92
Intensive care	–	–	–	–	–	228,23
Laboratory findings	–	167,50	186,96	208,42	236,24	287,56
Movement data	–	–	–	–	14,89	16,98
Radiological findings	–	–	–	–	–	0,97
Risk factors	–	–	–	–	0,03	0,05

As of July 2018, the ECRDW repository contains data from more than 2 million patients, more than 11 million diagnoses and more than 6 million cases with approximately 500 million data points.

The ECRDW currently collects administrative information such as demographic data, movement data, visit data, diagnoses (ICD-10GM), risk factors and severity of the disease from the SAP i.s.h. system. The EMS (SAP i.s.h.med) is used to load reports, findings and discharge letters. Metadata for biosamples is provided from the MySamples and CentraXX systems. Intensive care data originates from a legacy system (COPRA) as well as from the operative ICU system (m.life). The ECRDW receives information on findings from the laboratory via LIS (OPUS::L). Metadata on radiological examinations (including findings) are provided via a RIS (GE Centricity). Cardiovascular data, findings and values of cardiological echocardiographies as well as cardiac catheter examinations originate from another RIS (IntelliSpace CardioVascular). Depending on the source system, different times are therefore possible for the start of digital recording. The earliest capture times are shown in Table 2 analog to Table 1.

Table 2. Earliest time of recording for each domain

Domains	Minimal date (month/year)
Biospecimen	10/2012
Demographic data	09/1986
ICD diagnosis	01/2007
Intensive care	05/2005
Laboratory findings	06/2000
Movement data	04/2008
Radiological findings	12/2013
Risk factors	07/2007

3.2   Projects Implemented with ECRDW

The ECRDW of the MHH is productive since 07/2013 and provides data on research-relevant issues. In the period from 07/2013 to 07/2018, 48 project inquiries from 39 departments of the MHH were registered. Three project requests are annually recurring data deliveries (e.g. register implementation). Table 3 shows an overview of the registered projects in the period from 2013 to 2018 for the three central use cases (screening, epidemiological study, validation and data enrichment).

In some projects, text analysis methods were used to obtain additional features from full-text documents, such as radiological findings and discharge letters, and to make them available with data from the ECRDW's structured databases.

Researchers are able to use innovative methods, such as medical data mining, to identify new hypotheses about the amount of data due to the very large amount of data per project.

From a total of 33 MHH clinics, 16 clinics (48%) submitted an evaluation request to the ECRDW in 2018 (by July). In some projects, in addition to providing data, a screening was carried out in advance to check the feasibility of the project request. In relation to the provision of data for epidemiological study requests (31 projects) and data enrichment (16 projects), however, only in 9 projects a screening for patient data has been carried out.

Table 3. Number and nature of ECRDW-based research projects (2014–2018)

Year	# Projects	# Departments	Screening	Epidemiological study	Validation and data enrichment
2014	4	4		3	1
2015	4	4		1	3
2016	13	9	2	8	5
2017	8	6	2	6	2
2018	19	16	5	13	5
Total	48	39	9	31	16

3.3   Data and Process Quality

In order to ensure data and process quality, the process chain dimension "processing" from the developed PAR scheme was completely implemented in the ECRDW. The following criteria, among others, were taken into account:

Traceability of the Data Origin (Data Linage/Data Provenance): Additional columns (load date and update date) in the tables as reference to the source system of every record were added.

Traceability of the Loading Process: Errors that occur during load jobs can be assigned to an unique error table referencing additional tables providing information about job definition and run. Incorrect records remain in a staging table for each entity and are deleted only after successful loading in the core data warehouse.

Restart Capability: Integration jobs can be repeated at any time, since the data is only deleted from the staging area when the record is successful load in the data warehouse and the update of the data warehouse only takes place when the ETL process is complete and the temporary target tables are merged into the real tables.

Standardization of the Development of ETL Pipelines and Modeling: For the data integration we use three templates (Staging, Historization and Update) for the ETL processes, which are already predefined and only need to be adapted.

Referential Integrity: The artificially generated primary and foreign keys are based on adequate hash function.

Redundancy-Free: Duplicates are recognized between loading processes and within a loading process using a hash value. The script for managing duplicates is integrated into the standardized templates.

Performance: To optimize performance, lookup tables are created before the start of the loading processes (using the hash function for reference checks) and then truncated again.

Data Cleansing Scope: An own developed model for error codes is used which classifies errors at attribute level (or finer) and monitors them in an error reporting system for each subject area.

Time Variance: The changes of data over time (historization) are tracked via the concept of a temporal database.

Standardized reports generated on the basis of the error records are held in the staging area. These are automatically distributed to the ECRDW team from the source system (BI) so that the cause of the errors can be identified and eliminated. For the long-term analysis, the errors are classified as persistent and BI procedures are applied to these error tables to identify error constellations and proactively avoid them in the sense of datamining.

3.4   Data Protection and Security

In coordination with the data protection officer of the MHH, a data protection concept was developed that defines processes for data use and access. The data protection concept provides the storage of health data in pseudonymised form. The pseudonymisation of health data is a necessary step towards compliance with data protection in order to protect patients from the identification of their person. Instead of patient identifying data (IDAT/PID), pseudonyms (surrogate keys) are used. The pseudonyms are administered and assigned in the ECRDW.

The patient's consent to the use of his/her data for scientific purposes is registered with the MHH treatment contract when the patient is admitted. The current data protection concept stipulates that the patient's consent must be given for any data processing for scientific purposes. This is taken into account in every step of a data provision process.

To ensure data security, all ECRDW systems are backed up daily via a central backup concept. Security authentication and authentisation of ECRDW users takes place via the central MHH Active Directory. A transaction log archives all queries and executing users.

4   Discussion and Conclusion

The use of a data warehouse technology as the basis for the implementation of multidisciplinary data integration and analysis platform results in some significant advantages for clinical research, among others:

– Relief for the operative health care systems
– Support in the planning and implementation of studies
– Data enrichment of research databases with quality-assured information from central systems (e.g. laboratory systems, administrative systems, OR systems)
– Integration and storage of historic data repositories which are not usable for IT (legacy systems)

In addition, the research repository with consolidated data from different domains (HIS and research systems), offers a more complete data set and thus a basis for investigating relationships and potential patterns between disease progression and measures. The using of medical data mining methods based on the extensive and retrospective data of an ECRDW can serve as a valuable resource for the generation of innovative knowledge in all areas of medicine [23].

The development of a translational data integration and analysis platform is however a long-term process. Although the MHH ECRDW project was initiated in 2010 (by a core development team of two persons), it was made available for all health researchers in June 2013. The selection of the appropriate data warehouse development platform, data modelling, a development of an integration strategy, data protection, security and use and access concepts and finally testing, validation and maintenance phases are the time consuming but needed phases of an iterative development process. Another very important issue is the involvement of experts from different fields (computer scientists, physicians, statisticians, privacy protect officer, management etc.) in different phases of development.

Semantic modelling of clinical concepts as well as analysis of unstructured data and OMICs are further key issue. At the MHH, additional tools for text analysis based on Natural Language Processing are currently being developed for the scientific use of findings that are only available in (semi-)structured form (such as medical letters and findings).

The FAIR Data Principles published in 2016 [24] define fundamentals that research data and research data infrastructures must meet in order to ensure sustainability and reusability. As part of a research infrastructure, the ECRDW has been respecting these principles in some aspects since 2013:

- *Findable*: data and metadata machine-readable and searchable through central database management system
- *Accessible*: use and access concept; data provision in standard formats or standard interfaces (CSV, ODBC, HTTPS)
- *Interoperable*: structuring of data by standard vocabularies, classification systems (ICD, LOINC, OPS, etc.); metadata on semantics between data sets
- *Reusable*: metadata can be exported machine-readable

Due to the actuality of the FAIR Data Principles, the future development of the ECRDW should also take into account these principles. The developed PAR scheme complements this approach.

Even in the age of knowledge graphs, data integration remains a major challenge [25]. With the increasing number of source systems, complexity has always increased. Challenges in data preparation and harmonization still apply to new big data technologies as well. We expect that the use of a data warehouse-based solution as an already consolidated and plausibility checked source will probably lead to a simplification by merging with further data sources (e.g. when consolidating with web sources) [26].

Intense exchange and collaborations within the national projects and facilities is required to take advantage of synergy effects. The involvement of ECRDW is an important factor of the sustainability concept within the Data Integration Centres of the Medical Informatics Initiative [27] and further national and international projects (e.g. German Biobank Alliance [28] and EHR4CR [29]).

References

1. Meystre, S.M. et al.: Clinical Data Reuse or Secondary Use: Current Status and Potential Future Progress (2017). https://doi.org/10.15265/iy-2017-007
2. Tolxdorff, T., Puppe, F.: Klinisches data warehouse. Inform. Spektrum **39**, 233–237 (2016). https://doi.org/10.1007/s00287-016-0968-3
3. Strasser, N.: RDA - Systeme an Universitätskliniken in Deutschland (2010)
4. Teasdale, S., et al.: Secondary uses of clinical data in primary care. Inform. Prim. Care **15**(3), 157–166 (2007). https://doi.org/10.14236/jhi.v15i3.654

5. Bonney, W.: Applicability of business intelligence in electronic health record. Procedia - Soc. Behav. Sci. **73**, 257–262 (2013). https://doi.org/10.1016/j.sbspro.2013.02.050
6. Mucksch, H., Behme, W.: Das Data Warehouse-Konzept, 4th edn. Gabler Verlag, Wiesbaden (2000)
7. Gerbel, S., Laser, H., Haarbrandt, B.: Das Klinische Data Warehouse der Medizinischen Hochschule Hannover. In: Forum der Medizin_Dokumentation und Medizin_Informatik, no. 2, pp. 49–52 (2014)
8. Dugas, M., Lange, M., Müller-Tidow, C., Kirchhof, P., Prokosch, H.U.: Routine data from hospital information systems can support patient recruitment for clinical studies. Clin. Trials, 183–189 (2010). https://doi.org/10.1177/1740774510363013
9. Murphy, S.N., et al.: Serving the enterprise and beyond with informatics for integrating biology and the bedside (i2b2). J. Am. Med. Inform. Assoc. **17**, 124–130 (2010)
10. Jannot, A.-S., et al.: The Georges Pompidou University hospital clinical data warehouse: a 8-years follow-up experience. Int. J. Med. Inform. **102**, 21–28 (2017)
11. Medizinische Hochschule Hannover. Jahresbericht der MHH (2016). https://www.mh-hannover.de/fileadmin/mhh/bilder/ueberblick_service/publikationen/Jahresbericht_2016_GESAMT_FINAL_31.08.2017.pdf. Accessed 15 Sept 2018
12. OVH SAS: Understanding Tier 3 and Tier 4 (2018). https://www.ovh.com/world/dedicated-servers/understanding-t3-t4.xml. Accessed 15 Sept 2018
13. Gartner Inc.: Magic Quadrant for Analytics and Business Intelligence Platforms (2018). https://www.gartner.com/doc/reprints?id=1-4RVOBDE&ct=180226&st=sb. Accessed 15 Sept 2018
14. Inmon, W.H.: Building the Data Warehouse, 3rd edn. Ipsen, R. (ed.). Wiley, Hoboken (2002)
15. Arbeitsgruppe Erhebung und Nutzung von Sekundärdaten (AGENS), et al.: Deutsche Gesellschaft für Epidemiologie e. V (2012). http://dgepi.de/fileadmin/pdf/leitlinien/GPS_fassung3.pdf. Accessed 15 Sept 2018
16. Thadani, S.R., et al.: Electronic screening improves efficiency in clinical trial recruitment. J. Am. Med. Inform. Assoc. **16**(6), 869–873 (2009). https://doi.org/10.1197/jamia.M3119
17. Sandhu, E., et al.: Secondary uses of electronic health record data: benefits and barriers. Jt. Comm. J. Qual. Patient Saf. **38**, 34–40 (2012). https://doi.org/10.1016/S1553-7250(12)38005-7
18. Beresniak, A., et al.: Cost-benefit assessment of using electronic health records data for clinical research versus current practices: contribution of the electronic health records for clinical research (EHR4CR) European project. Contemp. Clin. Trials **46**, 85–91 (2016). https://doi.org/10.1016/j.cct.2015.11.011
19. Coorevits, P., et al.: Electronic health records: new opportunities for clinical research. J. Intern. Med. **274**, 547–560 (2013). https://doi.org/10.1111/joim.12119
20. Just, B.H., et al.: Why Patient matching is a challenge: research on master patient index (MPI) data discrepancies in key identifying fields. Perspectives in Health Information Management (2016). PMC4832129
21. Wang, R., Strong, D.: Beyond accuracy: what data quality means to data consumers. J. Manag. Inf. Syst. **12**, 5–33 (1997). https://doi.org/10.1080/07421222.1996.11518099
22. Rassmann, T.: Entwicklung eines Verfahrens zur integrierten Abbildung und Analyse der Qualität von Forschungsdaten in einem klinischen Datawarehouse, Dissertation (2018)
23. Prather, J.C., et al.: Medical data mining: knowledge discovery in a clinical data warehouse. In: Proceedings of the AMIA Annual Symposium, pp. 101–105 (1997). PMC2233405
24. Wilkinson, M.D., et al.: The FAIR guiding principles for scientific data management and stewardship. Sci. Data **3** (2016). https://doi.org/10.1038/sdata.2016.18

25. Konsortium HiGHmed: Heidelberg - Göttingen - Hannover Medical Informatics. http://www.medizininformatik-initiative.de/de/konsortien/highmed. Accessed 31 July 2018
26. German Biobank Alliance (GBA). https://www.bbmri.de/ueber-gbn/german-biobank-alliance/. Accessed 31 July 2018
27. Electronic Health Records for Clinical Research - (EHR4CR). http://www.ehr4cr.eu/. Accessed 31 July 2018

User-Driven Development of a Novel Molecular Tumor Board Support Tool

Marc Halfmann[1]([✉])(iD), Holger Stenzhorn[2](iD), Peter Gerjets[1],
Oliver Kohlbacher[2], and Uwe Oestermeier[1]

[1] Leibniz-Institut für Wissensmedien (IWM), Tübingen, Germany
m.halfmann@iwm-tuebingen.de

[2] Institute for Translational Bioinformatics, University Hospital Tübingen,
Tübingen, Germany

Abstract. Nowadays personalized medicine is of increasing importance, especially in the field of cancer therapy. More and more hospitals are conducting molecular tumor boards (MTBs) bringing together experts from various fields with different expertise to discuss patient cases taking into account genetic information from sequencing data. Yet, there is still a lack of tools to support collaborative exploration and decision making. To fill this gap, we developed a novel user interface to support MTBs. A task analysis of MTBs currently held at German hospitals showed, that there is less collaborative exploration during the meeting as expected, with a large part of the information search being done during the MTB preparation. Thus we designed our interface to support both situations, a single user preparing the MTB and the presentation of information and group discussion during the meeting.

Keywords: Personalized Medicine · Cancer therapy · Multitouch
Multiuser

1 Introduction

Personalized Medicine is a rapidly growing area in healthcare which fundamentally changes the way patients are treated. This is especially true for cancer therapy where more and more hospitals conduct molecular tumor boards (MTBs) bringing together experts from various clinical fields to jointly discuss individual patient cases [1]. Yet the exploration and discussion of the relevant data and information poses a tremendous challenge in this setting: Because of the experts' distinct background expertise and time constraints, all data and information need to be presented in a concise form that is easy to grasp and supports the consensual elaboration of sound treatment recommendations.

We approach this need by developing a software solution which focuses on a novel, intuitive graphical user interface that integrates and visualizes

This research has been funded by the German Federal Ministry of Education and Research within the project PersOnS (031L0030A) of the initiative i:DSem.

S. Auer and M.-E. Vidal (Eds.): DILS 2018, LNBI 11371, pp. 195–199, 2019.
https://doi.org/10.1007/978-3-030-06016-9_18

patient data from several clinical information, laboratory and imaging systems, etc., together with additional external information from cancer-research systems (e.g. cBioPortal), genomic and pathway databases (e.g. KEGG), as well as (bio)medical literature (e.g. PubMed). A task analysis of MTBs was conducted and clinical experts were repeatedly shown a current version of a prototypical interface to integrate their feedback into the development process.

2   Methods

To detect the actual needs towards such a software, we initially visited five German MTBs where we used a questionnaire and interviews to gather information about specific requirements of each MTB, management and workflows, perceived weaknesses and room for improvements, as well as ideas on what functionalities an ideal software tool should provide (for comparison see [2]).

The interviews were conducted with the MTB organizers to gain insight into their workflow during preparation of the board, the resources they use for research and the overall procedure of the meeting itself. In the questionnaires[1] we first asked about personal information, like the participant's role in the MTB or his/her expertise and then about the procedure of the MTB and included questions, like whether therapy successes/failures and comparable cases are discussed or if there is a demand for online research during the MTB. Participants could answer on a scale ranging from *not at all* or *never* to *yes* or *always*.

The resulting requirement analysis formed the basis of our development process which is now accompanied by regular visits to our local MTB to discuss whether our software indeed reflects the experts' specific needs and wishes. This allowed us to evaluate our interface design from the very beginning of the development and to adjust it to the actual needs of the MTB.

Our initial (naive) belief at project start was that during the MTB a group of experts would come together to jointly discuss and evaluate and data patient cases in depth on the spot. Hence our first prototype employed a large (84 inches) multi-touch screen table allowing the MTB participants to collaboratively and interactively explore all information sources and arrange the different kinds of documents, like clinical/patient data, radiologic images, research papers, etc. on the screen [3,4]. Our initial thought was that such a table would greatly improve collaborative decision making and reduce the cognitive load for the participants.

We implemented an initial prototype prior to completing the task analysis to show the capabilities of such an interface to some clinical experts. This was important, as based on their daily routine using standard clinical software solutions, they might lack the knowledge about the possibilities of novel multi-user interfaces. This also gave us important feedback of the actual needs for a MTB application. It is important to note, that clinical experts who were given the questionnaire were not shown the prototype prior to answering the questions.

---

[1] Questionnaire: https://bit.ly/persons-questionnaire.

3   Results

The interviews and our participation in the MTBs revealed that the given time constraints only allow for a relatively short discussion of the individual patient cases. As a consequence, there is no room for any interactive, collaborative research or data exploration and so basically all research needs to be done before.

We collected questionnaires from 23 participants. Although this seems a rather small number, it is important to note that at the time of our analysis only 5 hospitals in Germany were conducting MTBs with 15 to 25 participants per MTB. Thus our questionnaire was handed out to about 100 MTB participants making up for a feedback of almost 25%. After assigning each verbal scale numerical values from 1 = *never/not at all* to 4 = *yes/always*, our results showed that participants see only little demand for online research during the MTB (μ = 2.304) which is mostly due to the mentioned time constraints. Also therapy successes/failures ($\mu = 2.047$) or comparable cases ($\mu = 2.273$) are rather seldom discussed yet participants mentioned this as highly desirable ($\mu = 3.667$).

Based on these findings, i.e. time constraints during the MTB and large amount of work to be done in the preparation, we changed our initial idea of having one tool, that only supports the group situation in the MTB and we thus divided the interface into two parts: One for the clinical experts to use on their usual computers to perform their explorative research when preparing a MTB and for the MTB moderator to then integrate and (visually) condense the results into a succinct presentation, and the second one to provide the means to present this during the MTB on any touch-enabled screen dynamically.

The user interface focuses on the presentation layer, performs no data processing at all and is realized in HTML and JavaScript using Electron. The advantage of electron is not only its platform independence and high portability, it also features a very important element: webviews. One of the key aspects of our interface, as described in detail below, is the possibility to display different sources of information (webpages, documents, visualizations) for free arrangement and comparison, in a single- or multi-user scenario. While this is not achievable with pure HTML and javascript using iframes, electron webviews offer all the functionality needed for such an interface.

The interface is then securely connected by an encrypted REST interface with the back-end services located within the secured hospital network. Those services – implemented in Java with Spring Boot and its sub-frameworks – perform the actual data fetching, processing and integration from the source systems and provide further means for data protection and pseudonymization.

Figure 1 shows an example screenshot of the whole interface together with close-ups of the important elements. The development of the layout was guided by the following design principles derived from our task analysis: (1) Minimize the number of actions and interface elements to improve speed and simplicity of the workflow, (2) support memory offloading and information interpretation (e.g., by free arrangements and annotations of resources, simultaneous visibility of all information, near-hand processing of information [5]), and (3) use space in an optimal way to leave as much space as possible for visual data exploration.

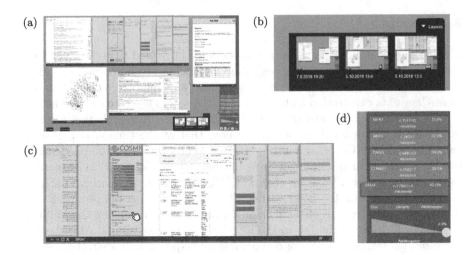

Fig. 1. Example screenshots of the prototypical interface: (a) complete interface, (b) collection of saved layout configurations with respective date, (c) webview containing multiple resources for comparison, (d) close-up of the gene sequencing data visualization (with activating and inactivating variants depicted in green and red respectively). For a more detailed overview of the interface (see footnote 4)

When opening a new case for the first time the user is shown a summary of the patient's clinical history including diagnosis, previous therapy, questions for the MTB and the potentially therapy relevant genetic variants. This summary is generated automatically and replaces the PowerPoint slides that were manually created by the MTB organizer in a time consuming process of copying text from the clinical information system or typing in by hand.

The interface itself is divided into three different areas. In the first area on the right side there is an overview of the therapy relevant variants (Fig. 1d), that can be filtered by allele frequency and sorted by various criteria. A color code indicates activating or inactivating variants. To get information for each variant, the current workflow requires the user to open up multiple browser tabs and manually search within the respective databases. To compare different sources of information for a single variant or even between multiple variants the user has to constantly switch between different tabs or applications. Our interface allows to open multiple sources of information (Fig. 1c) for one variant with a single click in the second area of the interface, the large exploration area that takes up most of the screen-space (Fig. 1a). From this overview single information sources can be opened in separate floating windows and which can be arranged and resized freely for exploration and comparison within that area. The user can also open information sources for different variants in parallel to compare them.

Creating the presentation for the MTB in the current preparation workflow requires the user to copy and paste all relevant information manually into a separate power point slide. In contrast, our interface allows to save an arrangement of information sources as a fixed layout without having to switch the application. Such a layout can later be restored for presentation during the meeting and due to the interactive properties of the HTML content these restored information sources can be used for further research and exploration as needed. Saved layouts are shown as thumbnails in a third area at the bottom (Fig. 1b).[2]

4   Conclusion and Outlook

We analyzed the workflow of MTBs currently held at German hospitals and developed a novel interface to support the preparation and the presentation for the meeting. In a next step, this interface will be evaluated in a clinical working environment and usability studies will be conducted. Also, additional features, deemed as valuable by clinical experts will be implemented as well, like the possibility to search for comparable local cases or the possibility for annotating documents and data to better document the decision making process.

References

1. van der Velden, D., et al.: Molecular tumor boards: current practice and future needs. Ann. Oncol. **28**, 3070–3075 (2017)
2. Hinderer, M., Boerries, M., Haller, F., Wagner, S., et al.: Supporting molecular tumor boards in molecular-guided decision-making - the current status of five German university hospitals. Stud. Health Technol. Inform. **236**, 48–54 (2017)
3. Morris, M., Huang, A., Paepcke, A., Winograd, T.: Cooperative gestures: multi-user gestural interactions for co-located groupware. In: Proceedings of the SIGCHI Conference on Human Factors in Computing Systems, pp. 1201–1210 (2006)
4. Streit, M., Schulz, H., Schmalstieg, D., Schumann, H.: Towards multi-user multi-level interaction. In: Proceedings of the Workshop on Collaborative Visualization on Interactive Surfaces, pp. 5–8 (2009)
5. Brucker, B., Brömme, R., Weber, S., Gerjets, P.: Learning on multi-touch devices: is directly touching dynamic visualizations helpful? In: 17th Biennial Conference of the European Association for Research on Learning and Instruction (EARLI), Tampere, Finland. Special Interest Group (SIG) 2 (2017)

---

[2] Interface Prototype Demo: https://bit.ly/persons-interface-prototype-demo.

Using Semantic Programming for Developing a Web Content Management System for Semantic Phenotype Data

Lars Vogt[1]([✉]) [iD], Roman Baum[1] [iD], Christian Köhler[1] [iD],
Sandra Meid[1] [iD], Björn Quast[2] [iD], and Peter Grobe[2] [iD]

[1] Universität Bonn, IEZ, An der Immenburg 1, 53121 Bonn, Germany
lars.m.vogt@googlemail.com
[2] Zoologisches Forschungsmuseum Alexander Koenig, Adenauerallee 160,
53113 Bonn, Germany

Abstract. We present a prototype of a semantic version of Morph·D·Base that is currently in development. It is based on SOCCOMAS, a semantic web content management system that is controlled by a set of source code ontologies together with a Java-based middleware and our Semantic Programming Ontology (SPrO). The middleware interprets the descriptions contained in the source code ontologies and dynamically decodes and executes them to produce the prototype. The Morph·D·Base prototype in turn allows the generation of instance-based semantic morphological descriptions through completing input forms. User input to these forms generates data in form of semantic graphs. We show with examples how the prototype has been described in the source code ontologies using SPrO and demonstrate live how the middleware interprets these descriptions and dynamically produces the application.

Keywords: Semantic programming · Phenotypic data · Linked open data
Semantic Morph·D·Base · Semantic annotation · Morphological data

1 Introduction

Ontologies are dictionaries that consist of labeled classes with definitions that are formulated in a highly formalized canonical syntax and standardized format (e.g. Web Ontology Language, OWL, serialized to the Resource Description Framework, RDF), with the goal to yield a lexical or taxonomic framework for knowledge representation [1]. Ontologies are often formulated in OWL and thus can be documented in the form of class-based semantic graphs[1]. Ontologies contain commonly accepted domain knowledge about specific kinds of entities and their properties and relations in form of classes defined through universal statements [2, 3], with each class possessing its own

---

[1] A semantic graph is a network of RDF/OWL-based triple statements, in which a given Uniform Resource Identifier (URI) takes the *Object* position in one triple and the *Subject* position in another triple. This way, several triples can be connected to form a semantic graph. Because information about individuals can be represented as a semantic graph as well, we distinguish class- and instance-based semantic graphs.

© Springer Nature Switzerland AG 2019
S. Auer and M.-E. Vidal (Eds.): DILS 2018, LNBI 11371, pp. 200–206, 2019.
https://doi.org/10.1007/978-3-030-06016-9_19

URI, through which it can be identified and individually referenced. Ontologies in this sense do not include statements about individual entities. Statements about individual entities are assertional statements. In an assertional statement individuals can be referred to through their own URI and their class affiliation can be specified by referencing this class' URI. If assertional statements are grounded in empirical knowledge that is based on observation and experimentation, we refer to them as empirical data. Empirical data can be formulated in OWL and thus documented in the form of instance-based semantic graphs. As a consequence, not every OWL file and not every semantic graph is an ontology—it is an ontology only if it limits itself to express universal statements about kinds of entities [3]. A knowledge base, in contrast, consists of a set of ontology classes that are populated with individuals and assertional statements about these individuals [3] (i.e. data). Ontologies do not represent knowledge bases, but are part of them and provide a means to structure them [4].

By providing a URI for each of their class resources, ontologies can be used to substantially increase semantic transparency and computer-parsability for all kinds of information. Respective URIs are commonly used for semantically enriching documents and annotating database contents to improve integration and interoperability of data, which is much needed in the age of Big Data, Linked-Open-Data and eScience [5–7]. Ontologies and their URIs also play an important role in making data maximally findable, accessible, interoperable and reusable (see FAIR guiding principle [8]) and in establishing eScience-compliant (meta)data standards [6, 7, 9–12].

An increasing number of organizations and institutions recognize the need to comply with the FAIR guiding principle and seek for technical solutions for efficiently managing the accessibility, usability, disseminability, integrity and security of their data. Content management systems in form of knowledge bases (i.e. Semantic web content management systems, S-WCMS) have the potential to provide a solution that meets both the requirements of organizations and institutions as well as of eScience.

Despite the obvious potential of ontologies and semantic technology in data and knowledge management, their application is usually restricted to annotating existing data in relational database applications. Although tuple stores that store information as RDF triple statements are capable of handling large volumes of triples and although semantic technology facilitates detailed data retrieval of RDF/OWL-based data through SPARQL [13] endpoints and inferencing over OWL-based data through semantic reasoners, not many content management systems have implemented ontologies to their full potential. We believe that this discrepancy can be explained by a lack of application development frameworks that are well integrated with RDF/OWL.

2 Semantic Programming

2.1 Semantic Programming Ontology (SPrO)

With SPrO [14] we extend the application of ontologies from providing URIs for annotating (meta)data and documenting data in form of semantic graphs stored and managed in a S-WCMS to using an ontology for software programming. We use SPrO like a programming language with which one can control a S-WCMS by describing it within a corresponding source code ontology. SPrO defines ontology resources in the form of classes, individuals and properties that the accompanying Java-based middleware interprets as a set of commands and variables. The commands are defined as annotation properties. Specific values and variable-carrying resources are defined as ontology individuals. Additional object properties are used to specify relations between resources, and data properties are used for specifying numerical values or literals for resources that describe the S-WCMS.

SPrO can be used to describe all features, workflows, database processes and functionalities of a particular S-WCMS, including its graphical user interface (GUI). The descriptions at their turn are contained in one or several source code ontologies in form of annotations of ontology classes and ontology individuals. Each annotation consists of a command followed by a value, index or resource and can be extended by axiom annotations and, in case of individuals, also property annotations. Contrary to other development frameworks that utilize ontologies (e.g. [15, 16]), you can use the resources of SPrO to describe a particular content management application within its corresponding source code ontology. The application is thus self-describing. The accompanying Java-based middleware decodes the descriptions as declarative specifications of the content management application, interprets them and dynamically executes them on the fly. We call this approach semantic programming.

2.2 Semantic Ontology-Controlled Application for Web Content Management Systems (SOCCOMAS)

SOCCOMAS [17] is a semantic web content management system that utilizes SPrO and its associated middleware. It consists of a basic source code ontology for SOC-COMAS itself (SC-Basic), which contains descriptions of features and workflows typically required by a S-WCMS, such as user administration with login and signup forms, user registration and login process, session management and user profiles, but also publication life-cycle processes for data entries (i.e. collections of assertional statements referring to a particular entity of a specific kind, like for instance a specimen) and automatic procedures for tracking user contributions, provenance and logging change-history for each editing step of any given version of a data entry. All data and metadata are recorded in RDF following established (meta)data standards using terms and their corresponding URIs from existing ontologies. Each S-WCMS run by SOC-COMAS provides human-readable output in form of HTML and CSS for browser requests and access to a SPARQL endpoint for machine-readable service requests. Moreover, it assigns a DOI to each published data entry and data entries are published under a creative commons license. When a data entry is published, it becomes openly and freely accessible through the Web. Hence, all data published by a S-WCMS run by SOCCOMAS reaches the five star rank of Tim Berners-Lee's rating system for Linked Open Data [18].

The descriptions of the features, processes, data views, HTML templates for input forms, specifications of input control and overall behavior of each input field of a particular S-WCMS are contained in its accompanying source code ontology, which is specifically customized to the needs of that particular S-WCMS. These descriptions also include specifications of the underlying data scheme that determines how user input triggers the generation of data-scheme-compliant triple statements and where these triples must be saved in the Jena tuple store in terms of named graph[2] and workspace (i.e. directory). For instance the morphological data repository semantic Morph·D·Base has its own source code ontology for its morphological description module (SC-MDB-MD [20]) that is specifically customized to the needs of semantic Morph·D·Base [19] (Fig. 1).

This way, the developers of semantic Morph·D·Base can use the general functionality that comes with SC-Basic and add upon that the features specifically required for semantic Morph·D·Base by describing them in SC-MDB-MD using the commands, values and variable-carrying resources from SPrO. After semantic Morph·D·Base goes online, its developers can still describe new input fields in SC-MDB-MD or new types of data entries in respective additional source code ontologies and therewith update semantic Morph·D·Base without having to program in other layers.

The application descriptions contained in SC-Basic and SC-MDB-MD organize the Jena tuple store into different workspaces, which at their turn are organized into different named graphs, each of which belongs to a particular class of named graphs. This enables differentially storing data belonging to a specific entry or version of an entry into different named graphs, which in turn allows for flexible and meaningful fragmentation of data and flexible definition of different data views.

2.3   Semantic Morph·D·Base as a Use-Case

Morphological data drive much of the research in life sciences [21, 22], but are usually still published as morphological descriptions in form of unstructured texts, which are not machine-parsable and often hidden behind a pay-wall. This not only impedes the overall findability and accessibility of morphological data. Due to the immanent semantic ambiguity of morphological terminology, researchers who are not experts of the described taxon will have substantial problems comprehending and interpreting the morphological descriptions (see *Linguistic Problem of Morphology* [23]). This semantic ambiguity substantially limits the interoperability and reusability of morphological data, with the consequence that morphological data usually do not comply with the FAIR guiding principles [8].

Semantic Morph·D·Base [19] enables users to generate highly standardized and formalized morphological descriptions in the form of assertional statements represented as instance-based semantic graphs. The main organizational backbone of a morphological description is a partonomy of all the anatomical parts and their sub-parts of the specimen the user wants to describe. Each such part possesses its own URI and is indicated to be an instance of a specific ontology class. Semantic Morph·D·Base allows reference to ontology classes from all anatomy ontologies available at BioPortal [24]. Parts can be further described (i) semantically through defined input forms, often referencing specific ontology classes from PATO [25], resulting in an instance-based

---

[2] A named graph identifies a set of triple statements by adding the URI of the named graph to each triple belonging to this named graph, thus turning the triple into a quad. The Jena tuple store can handle such quadruples. The use of named graphs enables partitioning data in an RDF store.

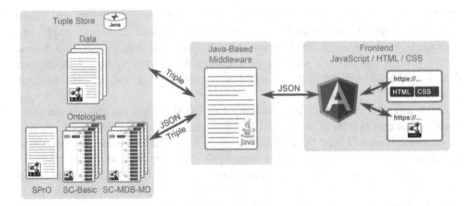

Fig. 1. Overall workflow of semantic Morph·D·Base [19] run by SOCCOMAS. Left: Jena tuple store containing the data of semantic Morph·D·Base as well as (i) the Semantic Programming Ontology (SPrO), which contains the commands, subcommands and variables used for describing semantic Morph·D·Base, (ii) the source code ontology for SOCCOMAS (SC-Basic), which contains the descriptions of general workflows and features that can be used by any S-WCMS, and (iii) the particular source code ontology for the morphological description module of semantic Morph·D·Base (SC-MDB-MD), which has been individually customized to contain the description of all features that are special to semantic Morph·D·Base. Middle: the Java-based middleware. Right: the frontend based on the JavaScript framework AngularJS with HTML and CSS output for browser requests and access to a SPARQL endpoint for machine requests.

semantic graph that we call a *Semantic Instance Anatomy* [26, 27], (ii) as semantically enriched free text, and (iii) through images with specified regions of interest which can be semantically annotated. All this information is stored in the tuple store and can be accessed through a web-based interface and a SPARQL endpoint. The *Semantic Instance Anatomy* graph is meaningfully fragmented into a sophisticated scheme of named graph resources, which additionally supports subsequent data retrieval and data analyses.

Because semantic Morph·D·Base is run by SOCCOMAS, all description entries not only possess their own unique URI, but also receive their own DOI when they are published and are freely and openly accessible through the Web. All data and metadata are stored as RDF triples in a Jena tuple store and can be searched using a SPARQL endpoint. Instances and classes referenced in these triples have their own globally unique and persistent identifiers and are findable through the endpoint. Both metadata as well as the descriptions themselves reference resources of well established ontologies, which substantially increases their interoperability and reusability. As a consequence, data and metadata in semantic Morph·D·Base comply with the FAIR principles.

Link to a live-demo of semantic Morph·D·Base: https://proto.morphdbase.de/

References

1. Smith, B.: Ontology. In: Floridi, L. (ed.) Blackwell Guide to the Philosophy of Computing and Information, pp. 155–166. Blackwell Publishing, Oxford (2003)
2. Schulz, S., Stenzhorn, H., Boeker, M., Smith, B.: Strengths and limitations of formal ontologies in the biomedical domain. RECIIS **3**, 31–45 (2009)
3. Schulz, S., Jansen, L.: Formal ontologies in biomedical knowledge representation. IMIA Yearb. Med. Inform. **2013**(8), 132–146 (2013)
4. Uschold, M., Gruninger, M.: Ontologies: principles, methods and applications. Knowl. Eng. Rev. **11**, 39–136 (1996)
5. Sansone, S.-A., Rocca-Serra, P., Tong, W., Fostel, J., Morrison, N., et al.: A strategy capitalizing on synergies: the reporting structure for biological investigation (RSBI) working group. OMICS: J Integr. Biol. **10**, 164–171 (2006)
6. Vogt, L.: The future role of bio-ontologies for developing a general data standard in biology: chance and challenge for zoo-morphology. Zoomorphology **128**, 201–217 (2009)
7. Vogt, L., Nickel, M., Jenner, R.A., Deans, A.R.: The need for data standards in zoomorphology. J. Morphol. **274**, 793–808 (2013)
8. Wilkinson, M.D., Dumontier, M., Aalbersberg, I.J., Appleton, G., Axton, M., et al.: The FAIR Guiding Principles for scientific data management and stewardship. Sci. Data **3**, 160018 (2016)
9. Brazma, A.: On the importance of standardisation in life sciences. Bioinformatics **17**, 113–114 (2001)
10. Brazma, A., Hingamp, P., Quackenbush, J., Sherlock, G., Spellman, P., et al.: Minimum information about a microarray experiment (MIAME)–toward standards for microarray data. Nat. Genet. **29**, 365–371 (2001)
11. Wang, X., Gorlitsky, R., Almeida, J.S.: From XML to RDF: how semantic web technologies will change the design of "omic" standards. Nat. Biotechnol. **23**, 1099–1103 (2005)
12. Vogt, L.: eScience and the need for data standards in the life sciences: in pursuit of objectivity rather than truth. Syst. Biodivers. **11**, 257–270 (2013)
13. SPARQL Query Language for RDF. W3C Recommendation, 15 January 2008
14. GitHub: code for semantic programming ontology (SPrO). https://github.com/SemanticProgramming/SPrO
15. Wenzel, K.: KOMMA: An application framework for ontology-based software systems. Semant. Web J. **swj89_0**, 1–10 (2010)
16. Buranarach, M., Supnithi, T., Thein, Y.M., Ruangrajitpakorn, T., Rattanasawad, T., et al.: OAM: an ontology application management framework for simplifying ontology-based semantic web application development. Int. J. Softw. Eng. Knowl. Eng. **26**, 115–145 (2016)
17. GitHub: code for semantic ontology-controlled web content management system (SOCCOMAS). https://github.com/SemanticProgramming/SOCCOMAS
18. Berners-Lee, T.: Linked data. (2009). https://www.w3.org/DesignIssues/LinkedData.html
19. Semantic Morph•D•Base Prototype. https://proto.morphdbase.de
20. GitHub: Code for semantic Morph·D·Base prototype. https://github.com/MorphDBase/MDB-prototype
21. Deans, A.R., Lewis, S.E., Huala, E., Anzaldo, S.S., Ashburner, M., et al.: Finding our way through phenotypes. PLoS Biol. **13**, e1002033 (2015)
22. Mikó, I., Deans, A.R.: Phenotypes in insect biodiversity research phenotype data : past and present. In: Foottit, R.G., Adler, P.H. (eds.) Insect Biodiversity: Science and Society, vol. II, pp. 789–800. Wiley, Hoboken (2018)

23. Vogt, L., Bartolomaeus, T., Giribet, G.: The linguistic problem of morphology: structure versus homology and the standardization of morphological data. Cladistics **26**, 301–325 (2010)
24. BioPortal. http://bioportal.bioontology.org/
25. Phenotype And Trait Ontology (PATO). http://obofoundry.org/ontology/pato.html
26. Vogt, L.: Assessing similarity: on homology, characters and the need for a semantic approach to non-evolutionary comparative homology. Cladistics **33**, 513–539 (2017)
27. Vogt, L.: Towards a semantic approach to numerical tree inference in phylogenetics. Cladistics **34**, 200–224 (2018)

Converting Alzheimer's Disease Map into a Heavyweight Ontology: A Formal Network to Integrate Data

Vincent Henry[1,2(✉)] , Ivan Moszer[2] , Olivier Dameron[3] ,
Marie-Claude Potier[2] , Martin Hofmann-Apitius[4] ,
and Olivier Colliot[1,2,5(✉)]

[1] Inria, Aramis Project-Team, Paris, France
vincent.henry@inria.fr
[2] ICM, Inserm U1127, CNRS UMR 7225, Sorbonne Université, Paris, France
olivier.colliot@upmc.fr
[3] Univ Rennes, CNRS, Inria, IRISA - UMR 6074, 35000 Rennes, France
[4] Fraunhofer SCAI, Sankt Augustin, Germany
[5] Department of Neurology and Neuroradiology,
AP-HP, Pitié-Salpêtrière Hospital, Paris, France

Abstract. Alzheimer's disease (AD) pathophysiology is still imperfectly understood and current paradigms have not led to curative outcome. Omics technologies offer great promises for improving our understanding and generating new hypotheses. However, integration and interpretation of such data pose major challenges, calling for adequate knowledge models. AlzPathway is a disease map that gives a detailed and broad account of AD pathophysiology. However, AlzPathway lacks formalism, which can lead to ambiguity and misinterpretation. Ontologies are an adequate framework to overcome this limitation, through their axiomatic definitions and logical reasoning properties. We introduce the AD Map Ontology (ADMO), an ontological upper model based on systems biology terms. We then propose to convert AlzPathway into an ontology and to integrate it into ADMO. We demonstrate that it allows one to deal with issues related to redundancy, naming, consistency, process classification and pathway relationships. Further, it opens opportunities to expand the model using elements from other resources, such as generic pathways from Reactome or clinical features contained in the ADO (AD Ontology). A version of ADMO is freely available at http://bioportal.bioontology.org/ontologies/ADMO.

Keywords: Alzheimer's disease · Ontology · Disease map
Model consistency

1 Introduction

Alzheimer's disease (AD) is a progressive neurodegenerative disorder of the brain that was first described in 1906. The intense activity of AD research constantly generates new data and knowledge on AD-specific molecular and cellular processes (a Medline search for "Alzheimer disease" results in over 135,000 articles, as of June 30, 2018).

© Springer Nature Switzerland AG 2019
S. Auer and M.-E. Vidal (Eds.): DILS 2018, LNBI 11371, pp. 207–215, 2019.
https://doi.org/10.1007/978-3-030-06016-9_20

However, the complexity of AD pathophysiology is still imperfectly understood [1]. These 110 years of efforts have essentially resulted in one dominant paradigm to underline the causes of AD: the amyloid cascade [2]. Therapeutics targeting this pathway failed to lead to curative outcome for humans, strongly suggesting the need for alternative hypotheses about AD etiology.

Since the turn of the century, omics technologies lead to a more comprehensive characterization of biological systems and diseases. The production of omics data in AD research opens promising perspectives to identify alternatives to the amyloid cascade paradigm. The current challenge is thus to integrate these data in an appropriate way, in order to propose new hypotheses and models about AD pathophysiology.

Systems medicine disease maps (DM) provide curated and integrated knowledge on pathophysiology of disorders at the molecular and phenotypic levels, which is adapted to the diversity of omics measurements [3, 4, 5]. Based on a systemic approach, they describe all biological physical entities (i.e. gene, mRNA, protein, metabolite) in their different states (e.g. phosphorylated protein, molecular complex, degraded molecule) and the interactions between them [6]. Their relations are represented as biochemical reactions organized in pathways, which encode the transition between participants' states as processes. AlzPathway is a DM developed for AD [3]. It describes 1,347 biological physical entities, 129 phenotypes, 1,070 biochemical reactions and 26 pathways.

The information contained in DM is stored in syntactic formats developed for systems biology: the Systems Biology Graphical Notation (SBGN) [7] and the Systems Biology Markup Language (SBML) [8]. While syntactic formats are able to index information, they are not expressive enough to define explicit relationships and formal descriptions, leading to possible ambiguities and misinterpretations. For AlzPathway, this defect in expressiveness results in the lack of formalism and thus of: (a) hierarchy and disjunction between species (e.g. between "Protein" and "Truncated Protein" or between "Protein" and "RNA", respectively), (b) formal definition of entities (such as phenotypes), (c) formal relationships between reactions and pathways (that are missing or are managed as cell compartments), (d) uniformity of entities' naming (e.g. complexes that are labelled by their molecular components or by a common name) and (e) consistency between reactions and their participants (e.g. translation of genes instead of transcripts).

Compared to syntactic formats, the Web Ontology Language (OWL), a semantic format used in ontologies, has higher expressiveness [9] and was designed to support integration. It is thus a good candidate to overcome the previous limitations.

An ontology is an explicit specification of a set of concepts and their relationships represented in a knowledge graph in semantic format. Ontologies provide a formal naming and definition of the types (i.e. the classes), properties, and interrelationships between entities that exist for a particular domain. Moreover, knowledge and data managed by an ontology benefit from its logical semantics and axiomatic properties (e.g. subsumption, disjunction, cardinality), which supports automatic control of consistency, automated enrichment of knowledge properties and complex query abilities [10].

The Alzheimer's Disease Ontology (ADO) [11] is the first ontology specific to the AD domain. ADO organizes information describing clinical, experimental and molecular features in OWL format. However, the description of the biological systems of ADO is less specific than that of AlzPathway.

Considering that (1) semantic formats can embed syntactic information, (2) DM provide an integrative view adapted to omics data management and (3) an ontological model is appropriate to finely manage data, the conversion of AlzPathway into a formal ontology would bring several assets, including an efficient integration of biomedical data for AD research, interconnection with ADO and an increased satisfiability of the resources.

We propose the Alzheimer Disease Map Ontology (ADMO), an ontological upper model able to embed the AlzPathway DM. Section 2 is devoted to the description of the ADMO model. In Sect. 3, we describe a method to convert AlzPathway in OWL and how ADMO can manage the converted AlzPathway and automatically enhance its formalism. Section 4 presents elements of discussion and perspectives.

2 Ontological Upper Model: Alzheimer Disease Map Ontology

The initial definition of an ontological model aims to design a knowledge graph that will drive its content. In a formal ontology, the relationships are not only links between classes, but also constraints that are inherited by all their descendants (subclasses). Thus, the choices of axioms that support high level classes and their properties are key elements for the utility of the model.

The Systems Biology Ontology (SBO) [12] is a terminology that provides a set of classes commonly used to index information in SBML format. These classes conceptualize biological entities at an adequate level of genericity and accuracy that supports a wide coverage with few classes and enough discrimination. We selected a set of 54 SBO terms from "process" or "material entity" for reactions and molecules as a first resource of subclasses of processes and participants, respectively. The modified Edinburg Pathway Notation (mEPN) [13] is another syntactic format based on systems approach. Its components provide a refined set of molecular states that complete the SBO class set. Following class selection from SBO and mEPN, we designed a class hierarchy between them. We systematically added disjointness constraints between the generic sibling subclasses of participants in order to ensure that process participants belong to only one set (e.g. a gene cannot be a protein and reciprocally). We did not apply the same rule to the processes' subclasses as a reaction may refer to different processes (e.g. a transfer is an addition and a removal).

Properties consistent with a systems approach (i.e. *part_of, component_of, component_process_of, has_participant, has_input, has_output, has_active_participant, derives_from* and their respective inverse properties) were defined from the upper-level Relation Ontology (RO) [14]. Then, we formally defined our set of classes with these properties and cardinalities to link processes and participants with description logic in SHIQ expressivity (e.g. a transcription has at least one gene as input and has at least one mRNA as output; a protein complex formation has at least two proteins as input and has at least one protein complex as output).

The design of the ADMO upper ontological model based on SBO, mEPN, RO and personal addition resulted in 140 classes (42 processes' subclasses and 83 participants subclasses) and 11 properties formally defined by 188 logical axioms in description logic (Fig. 1). This model is based on a simple pattern as our knowledge graph involves only three types of properties: (1) the *is_a* (*subclass_of*) standard property, (2) the *has_part* standard property and its sub-properties *has_component* and *has_component_process* and (3) the *has_participant* property and its sub-properties *has_input*, *has_output* and *has_active_participant*.

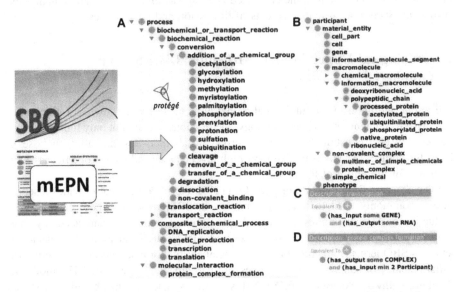

Fig. 1. Alzheimer disease map ontology model design. Classes were extracted from the Systems Biology Ontology (SBO) and the modified Edinburg Pathway Notation (mEPN) into Protégé. Classes were hierarchized as subclasses of process (A) or participant (B). Using properties from the Relation Ontology (RO), classes were formally defined in description logic, as illustrated in the case of transcription (C) and protein complex formation (D) processes.

3   AlzPathway Conversion and Integration into ADMO

AlzPathway elements were extracted and stored in a structured table using home-made Python scripts. In this table, each biological entity was indexed by one of the high-level participants' subclasses of ADMO and all processes were in correspondence with their participants. The table also contains class annotations such as the AlzPathway identifier (ID), and IDs from other knowledge bases such as UniProt [15] for participants and KEGG [16] for processes. The table is structured to integrate component information for multiplex entities (e.g. protein complex) and location information for the process (e.g. cell type or cell part). The table was then manually curated as described below.

In AlzPathway, native and modified proteins (e.g. phosphorylated or activated) may have the same label and same Id. In order to specify these different states, we added a suffix to modified protein labels (e.g. "_P" or "_a" for phosphorylated or activated, respectively).

In AlzPathway, phenotypes are participants. But several of them are named with a process name, pathway label or molecule type (e.g. microglial activation, apoptosis or cytokines, respectively). In order to deal with these ambiguities, 26 phenotypes were reclassified as molecules (e.g. cytokine) or cellular components (e.g. membrane) and 14 names that referred to processes or pathways were changed into processes' participant names (e.g. apoptosis became apoptotic signal). In addition, 5 phenotypes that were named with a relevant pathway name (e.g. apoptosis) were added to the initial set of the 26 AlzPathway's pathways.

AlzPathway only describes a subset of genes, mRNA and proteins. As omics technology can capture data at the genome, transcriptome or proteome levels, we added missing information in order to complete some correspondences between genes and gene products. This resulted in the addition of 406 genes, 415 mRNA and 194 proteins and protein complex states.

Then, using the ontology editor Protégé, the content of the structured table was imported into ADMO using the Protégé Cellfie plugin. Entities information were integrated as subclasses of ADMO participants classes. During the integration, we also added a new property *has_template* (sub-property of *derives_from*) to formally link a gene to its related mRNA and a mRNA to its related protein. Reactions were integrated as independent subclasses of the "process" class. Then, automated reasoning was used to classify them as subclasses of the ADMO upper model process classes depending on their formal definition (see Fig. 2a*). The 1,065 inferred *subclass_of* axioms corresponding to this refined classification of processes were then edited. During their import, process classes from AlzPathway were formally linked to their respective location through the RO property: *occurs_in*.

While AlzPathway does not formally link pathways and their related biochemical reactions, pathways were manually imported. For each pathway, a class "reaction involved in pathway *x*" was created and defined both as "reaction that *has_participant* the molecules of interest in *x*" and "*component_process_of* pathway *x*". For example, the class "reaction involved in WNT signaling pathway" *has_participant* "WNT" and is a *component_process_of* "WNT signaling pathway". Then, using automated reasoning, all reactions having participants involved in pathway *x* were classified as subclasses of "*component_process_of* pathway *x*" classes and were linked to the pathway with the *component_process_of* property by subsumption. For example, "SFRP-WNT association" is automatically classified as subclass of "reaction involved in WNT signaling pathway" (see Fig. 2b*) and inherits from its properties *component_ process_of* "WNT signaling pathway" (see Fig. 2b**). The 355 inferred *subclass_of* axioms corresponding to reactions involved in one of the 22 pathways were then edited.

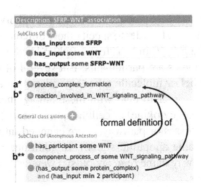

Fig. 2. Example of automated reasoning on Protégé. Asserted axioms are in uncoloured lines and inferred axioms are highlighted in yellow. Following automated reasoning, SFRP-WNT heterodimer association is classified as subclass of "protein complex formation" (a*) and of "reaction involved in WNT signaling pathway" classes (b*), thus it inherits of the *component_process_of* "WNT_signaling pathway" property (b**).

As a result, ADMO embeds AlzPathway in a consistent network containing 2,132 classes (2,175 disjoint participants, including 88 phenotypes or signals, 1,038 disjoint processes and 22 pathways) in relation with 10,964 logical axioms before and 12,373 logical axioms after automated reasoning, respectively. Specific efforts were dedicated to the design of classes hierarchy and formal definition with description logic axioms, leading to explicit relations between processes and biological entities. These axioms were inherited by classes imported from AlzPathway, resulting in the formal and precise description of the elements of AD pathophysiology. Thus, following automated reasoning, only 21 out of 643 AlzPathway's reactions generically classified as "transition" or "unknown transition" remained unaffected to a specific process of the ADMO upper model, such as metabolic reaction, phosphorylation or activation. Moreover, mis-affected processes were consistently affected to a specific process (e.g. translation instead of transcription). In addition, 355 reactions were formally defined as subprocess of pathways of interest.

The conversion of AlzPathway also benefits from the ADMO simple pattern of relationships (Fig. 3) in which new properties were added: the *derives_from* property that links a modified protein to its native form, the *has_template* property that links a native gene product to its mRNA and gene, and the *occurs_in* property that links a process to its cellular location.

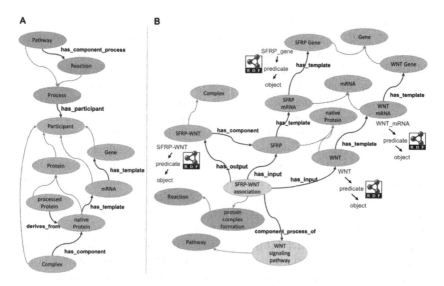

Fig. 3. Alzheimer disease map ontology (ADMO) pattern (A) and application to AlzPathway (B). AlzPathway classes (B; illustrated for the SFRP-WNT association process ant its participants) are now subclasses of ADMO classes (A). Each class of AlzPathway may be instantiated by the corresponding entities as individuals. Then, entities can be related to different objects in an RDF schema such as patients and experiments, or more specifically to values such as SNP for genes, relative expression for mRNA, and concentration for proteins.

4 Discussion

We proposed the ADMO ontological model in order to manage the conversion and integration of AlzPathway in OWL format. By converting AlzPathway into an OWL ontology, we increased its formalism. All entities are now formally defined and interconnected within a consistent network. While AlzPathway contained several ambiguities, our efforts on formalism at a semantic level for phenotypes and description logic in ADMO classes allowed us to solve inconsistencies. Moreover, the combination of SBO and mEPN provided a more precise specification of processes and biological entity states within the system compared to SBML or SBGN, which was beneficial for the specification of AlzPathway reactions following its import into ADMO.

Unlike DM, ontologies are not adapted for graphical visualization but present a higher flexibility to integrate new elements in the knowledge graph, as we did by adding 865 genes and mRNA. Moreover, during the conversion step, AlzPathway's internal IDs were retained as class annotations, allowing interoperability between the initial and converted AlzPathway. Taking advantage of the knowledge graph and its semantic links, the ID information are retrievable from a derived molecule to its native form following the *derives_from* or *has_component* properties that link each of these classes.

Furthermore, the increased formalism requires to assert a participant as subclass of the most representative class and thus, clarifies the status of the entities. In several standard bioinformatics knowledge resources (e.g. UnitProt [15], KEGG [16]), a same ID refers to a gene or a protein and *in fine* to a set of information, such as gene, interaction, regulation and post translation modification (PTM), which are thus not specifically discriminated. However, current omics technologies are able to generate data focused on specific elements of the systems (gene mutation, relative gene expression, protein concentration...). This is underexploited by standard resources. Based on DM approaches, we provided an ontology that (a) represents the complexity of a system such as AD pathophysiology and (b) is designed to specifically integrate each type of omics data as an instance of the explicit corresponding class.

The next possible step is to instantiate the model with biomedical omics data. To this end, the Resource Description Framework (RDF) semantic format is appropriate as it was specifically designed for representing a knowledge graph as a set of triples containing directed edges (semantic predicates). Different RDF schemas were already developed in the field of molecular biology (BioPax [17]) or more specifically for AD biomedical research (neuroRDF [18]). The Global Data Sharing in Alzheimer Disease Research initiative [19] is also a relevant resource to help find appropriate predicates to enrich RDF schemas and refine subject information (age, gender, clinical visit...). Depending on the need of a given study, users may design RDF schemas with their own predicates of interest. Then, this RDF schema can be integrated in our ontology by adding data as instances of its corresponding specific classes (Fig. 3B). Therefore, instantiation opens perspectives for complex querying, both richer and more precise than indexing.

DM are based on systems biology approaches, allowing one to take each part of the system into consideration. Our ontology goes one step further by formally defining the different elements of the system and linking them with the biochemical reaction and pathway levels. Here, we relied on AlzPathway, but additional resources could be used, such as Reactome [4] which provides a wide range of generic curated human biochemical reactions and pathways. Our ADMO upper ontological model provides an interesting framework to embed generic resources and thus harmonize AlzPathway and those resources. By converting and integrating AlzPathway in OWL format, the resulting ontology is ready to be connected with ADO and its clinical knowledge description. Owing to its specificity on biochemical reactions, an interoperable and formal version of AlzPathway should be a relevant complement to ADO. This offers new avenues for increasing the scale of representation of AD pathophysiology in our framework. In the same way, the genericity of processes and participants described in the ADMO upper model opens the perspective to harmonize specific DM from different neurodegenerative disorders such as the Parkinson's disease map [5] and others.

Acknowledgements. The research leading to these results has received funding from the program "Investissements d'avenir" ANR-10-IAIHU-06 (Agence Nationale de la Recherche-10-IA Institut Hospitalo-Universitaire-6) and from the Inria Project Lab Program (project Neuromarkers).

References

1. 2018 Alzheimer's disease facts and figures. Alz. Dem. **14**(3), 367–429 (2018)
2. Golde, T.E., Schneider, L.S., Koo, E.H.: Anti-aβ therapeutics in Alzheimer's disease: the need for a paradigm shift. Neuron **69**(2), 203–213 (2011)
3. Ogishima, S., et al.: AlzPathway, an updated map of curated signaling pathways: towards deciphering alzheimer's disease pathogenesis. Met. Mol. Biol. **1303**, 423–432 (2016)
4. Fabregat, et al.: The reactome pathway knowledgebase. NAR **46**(D1), D649–D655 (2018)
5. Fujita, K.A., et al.: Integrating pathways of Parkinson's disease in a molecular interaction map. Mol. Neurobiol. **49**(1), 88–102 (2014)
6. Kitano, H., et al.: Using process diagrams for the graphical representation of biological networks. Nat. Biotech. **23**(8), 961–966 (2005)
7. Mi, H., et al.: Systems Biology graphical notation: activity flow language level 1 version 1.2. J. Integr. Bioinf. **12**(2), 265 (2015)
8. Smith, L.P., et al.: SBML level 3 package: hierarchical model composition, version 1 release 3. J. Integr. Bioinf. **12**(2), 603–659 (2015)
9. Schaffert, S., Gruber, A., Westenthaler, R.: A semantic wiki for collaborative knowledge formation (2005)
10. Mizoguchi, R.: Tutorial on ontological engineering. New. Gen. Comp. **21**(4), 363–364 (2003)
11. Malhotra, et al.: ADO: a disease ontology representing the domain knowledge specific to Alzheimer's disease. Alz. Dem. **10**(2), 238–246 (2014)
12. Courtot, M., et al.: Controlled vocabularies and semantics in systems biology. Mol. Syst. Biol. **7**, 543 (2011)
13. Freeman, T.C., et al.: The mEPN scheme: an intuitive and flexible graphical system for rendering biological pathways. BMC Syst. Bio. **4**, 65 (2010)
14. Smith, B., et al.: Relations in biomedical ontologies. Gen. Biol. **6**(5), R46 (2005)
15. The UniProt Consortium: UniProt: the universal protein knowledgebase. NAR 46(5), 2699 (2018)
16. Kanehisa, M., et al.: KEGG: new perspectives on genomes, pathways, diseases and drugs. NAR **45**(D1), D353–D361 (2017)
17. Demir, E., et al.: BioPAX – a community standard for pathway data sharing. Nat. Biotech. **28**(9), 935–942 (2010)
18. Iyappan, et al.: NeuroRDF: semantic integration of highly curated data to prioritize biomarker candidates in Alzheimer's disease. J. Biomed. Sem. **7**, 45 (2016)
19. Ashish, N., Bhatt, P., Toga, A.W.: Global data sharing in Alzheimer disease research. Alzheimer Dis. Assoc. Disord. **30**(2), 160–168 (2016)

Author Index

Printed in the United States
By Bookmasters